GUIDE TO AMERICAN VINEYARDS

"This guide offers a compelling look at the wineries—both large and small—of America. At each vineyard you get a good taste of the history and high points of the place, its vintners philosophy, and its wines. You can sip at this book from your own armchair and learn a lot about America's growing wine culture, or you can get on the road with it and use it as your guide."

—K.M. Kostyal, Contributing Editor, *National Geographic Traveler*

"A user-friendly guide from someone who obviously compiled the information herself. . . . Many states you don't think of having wineries are included."

—Jerry D. Mead, Editor & Publisher, *The Wine Trader*

"Much more than a mere directory . . . A Guide to American Vineyards *. . . makes for good reading, with historical anecdotes and winery tips and notes.*

—Richard L. Elia, Publisher, *The Quarterly Review of Wines*

"An enlightened and easy-to-read guide for tourist or taster."

—Darryl M. Roberts, Wines International

Guide to American Vineyards

A GUIDE TO THE BEST WINERIES
FOR TOURING AND TASTING

by
Pamela Stovall

A Voyager Book

Old Saybrook, Connecticut

Text illustrations by Duane Perreault
Wine labels courtesy of the wineries

Library of Congress Cataloging-in-Publication Data

Stovall, Pamela.
Guide to American vineyards : a guide to the best wineries for touring and tasting / by Pamela Stovall. — 1st ed.
p. cm.
Includes index.
ISBN 1-56440-057-3
1. Wine and wine making—United States. I. Title
TP557.S763 1992
641.2′2′ 0973—dc20 92-12209
CIP

Manufactured in the United States of America
First Edition/First Printing

To Dorie Stovall, who started as my mother, then became my sister,
and finally my friend. To Howard M. Stovall,
even though it's just a book.

And to Rick, for all the reasons he already knows.

Contents

Acknowledgments

To the people who offered shelter from the road (and a chance for me to get out of the back of my pickup truck): Naomi and Bub, Austin, Texas; Vickie, Austin, Texas; Bob and Tino, Nogales, Arizona; John and Martha (in absentia), San Diego, California; Lorenzo, Los Angeles, California; Michele and Craig, Carmel, California; Mr. Ed, Gresham, Oregon; Marlin, Salt Lake City, Utah; Alice and Ken, Denver, Colorado; Mark and Lynne, Red Wing, Colorado; Russ, Santa Fe, New Mexico; Vic, Altus, Arkansas; Pepper, Madison, Wisconsin; Audrey and Howard, Chicago, Illinois; Rick, Munchie Mansion; Dorie, Howard, and Dennis, Grand Rapids, Michigan; Debbie, Phil, Emily, and Sarah, Comstock Park, Michigan; Kim (PMS Construction) and Donna Jean, Grand Rapids, Michigan; Louise, Jack, and Karl, Washington, D.C.; Joyce, Suiling, Julie, and Al, Lovingston, Virginia; Carl and Larry, La Te Da, Ft. Lauderdale, Florida.

I would like to thank all the wineries that graciously answered my questionnaires and that play pleasant hosts to those interested in wine. Special thanks to all the staff and owners that took time to talk with me.

I would also like to thank Eleanor, who helped get me into this mess; Sandra Choron of March Tenth, who would never give up; and Chico, *siempre conmigo.*

Introduction

Whether you're a wine aficionado or someone merely looking for something fun to do, you couldn't pick a better place to visit than a winery. Wineries offer something for everyone. Visiting wineries and sampling wine make a great day trip, weekend diversion, or year-round adventure. In the last twenty years the number of states with wineries has almost doubled. The staff at each winery included in this guide is waiting to pour you some wine or offer a tour of the facilities.

Each winery has a personality of its own. One may be a large corporate investment with slick videos and tour guides in uniforms, another a down-home family business offering visitors an impromptu lunch in the winemaker's kitchen. Their appeal has to do with more than just wine; it's the owner's or winemaker's personality that makes each one unique and makes the visit a special experience.

Year-round, there's always something going on at a winery. In the winter many wineries have cross-country ski trails and offer some sort of hot spiced wine. Vines literally explode in the spring with buds that will soon turn to grapes. In the summer you can picnic in the vineyards or other scenic spots on the winery's grounds. The fall brings harvest time and a great opportunity to see the winemakers in action.

There are more than a thousand wineries in the United States, but the ones included in this book were selected for the traveler looking for an enjoyable experience. I began my search for enjoyable wineries with a directory of every licensed winery in the United States. Then, based on the premise that a visit to a winery should include a walk among the vines, I found out which grow their own grapes (some wineries buy the grapes or the juice from others).

The list was further reduced by selecting only wineries that cater to visitors and offer tours and tastings. Some were also selected if they offered something unique, such as a tram ride to the winery. I focused on establishments that had been in operation for a few years and seemed to be on good financial footing and therefore likely to be doing business when you come to visit.

Then I hit the road for almost a year, traveling coast to coast, to personally search out the best wineries. Sometimes I found incredible wineries that didn't meet all the criteria for inclusion; in these cases exceptions were made. I also discovered that some wineries look good on paper, but when I stopped by, I found that they were not open on the days advertised or did not treat visitors well.

Whenever possible I did not advertise the fact that I was working on a guide. I wanted to see how you, the reader, would be treated during your visit. Sometimes, though, this was not possible. If I walked into a winery and was the only visitor there, eventually someone would ask me why I was taking notes, and of course I would tell them.

I talked with the people who were working at the wineries rather than interview the owners. My goal was to give an accurate portrayal of what and whom you would find during your visit.

While taking the tours and tasting the wines, I discovered that not all wineries are created equal—thank God. You can't compare a winery in Alabama with one in California, but each has something special to offer. Of course, some states, such as California and New York, have enough wineries for a book of their own. In these cases I provided a sample of the best the state has to offer.

All tours and tastings are free unless noted. Also, wineries that are especially equipped to handle children have been noted as such.

I'm sure you'll have as much fun visiting and reading about the wineries as I had writing about them. Enjoy.

Wine in the United States

The earliest explorers of the North American continent found wild grapes growing in abundance. The French Huguenots in the South made wine from these wild grapes—probably Muscadines, *Vitis rotundifolia*—as early as the 1560s. This was truly America's first wine. The grapes were found throughout the South from Florida to Tennessee.

In the Northeast, colonists made wine from *Vitis labrusca,* the wild grapes found in that region. But the labrusca has a strong, distinct flavor, called "foxy." The colonists were used to the wine made from European grapes. Immigrants from France, Italy, Spain, and other areas had previously brought their vines to America. Unfortunately these vines, *Vitis vinifera,* did poorly on the East Coast. So while the colonists experimented with vinifera, they continued making wine with labrusca.

During the 1700s and 1800s, wineries sprang up in almost every state on the East Coast and in the Midwest. Native varieties were most often used. In the Northeast the Concord was king. New York, Ohio, and Missouri were producing millions of gallons of wine and were winning international awards with such grapes as Niagara, Delaware, and Catawba.

Out in the West, from the 1600s to the 1700s, Franciscan priests traveled north from Mexico, bringing with them the Mission grape. At each mission they would plant a vineyard and use the grapes to make sacramental and medicinal wine. They passed through Arizona, New Mexico, and California. By the late 1700s the Mission grape was traveling up the coast of California. The missions spread in the state from San Diego north to Sonoma. By 1823 almost thirty missions had been started. Within a few years commercial wine production began.

Just as many wineries in various states were hitting their stride in the late 1800s, problems began. Some vineyards on the East Coast were shut down during the Civil War, either from lack of manpower to care for the grapes or because the vineyards had been turned into battlefields.

Then in the late 1800s the root louse phylloxera, which had spread in Europe, crossed the Atlantic and infected American vinifera vines. A solution was found when viticulturists realized that some native American varieties were immune. To save the great vineyards of Europe, native American rootstock was sent overseas. French vines were grafted onto the resistant labrusca rootstock. Vinifera vines in the United States were also grafted onto native American varieties. But just as one solution was found, another problem came along—Prohibition. Many winery owners found themselves in "dry" states—ones that banned alcohol consumption and production—long before Prohibition shut down the whole country.

The 18th Amendment in 1919 and the Volstead Act of 1920 made illegal "the manufacture, sale, or transportation of intoxicating liquors . . . for beverage purposes." With Prohibition, most wineries tore out their vineyards and destroyed their equipment, although some continued to produce grape juice, medicinal wine, or sacramental wine.

With the repeal of Prohibition in 1933, few wineries were prepared to begin production again right away. It takes years for vineyards to mature, and most wineries had torn theirs out. They therefore had no grapes. Also, the wine industry in many states never recovered because state laws against alcohol production and consumption remained intact. New wineries found it difficult to open legally.

Within the last thirty years, however, there's been an explosion in the American wine industry. Renewed interest in American wines has helped the growth of wineries in many states, and the number of states with wineries has almost doubled. All states were helped when California wines began consistently winning top awards and international attention. It helped Americans focus attention on wine produced in their own country.

Several other developments also aided the U.S. wine industry. New research with hardier varieties has made it possible to grow wine grapes in states previously considered too cold for vineyards. French-American hybrids, hardier than the vinifera, have spread from the East Coast into the Midwest. New growing techniques have improved production. And state bills legalizing farm wineries have led to the establishment of many such operations. It has only been in the last ten to twenty years in many states that a winery owner could offer samples and sell wine at the winery.

Today, then, American wine drinkers are in an enviable position. They have the opportunity to try wine made from three families of grapes, the native American, the French-American hybrids, and the Old World vinifera. And it's all in their own backyard.

From Grape to Glass: How Wine Is Made

Although there are some differences among wineries in how they grow their grapes and make their wine, the overall process is similar. The vine carries nutrients and moisture from the soil to the grape. If the vine receives the proper nutrients and adequate moisture and has enough sun, the vine will produce a quality grape. And possibly the grape will be made into a quality wine.

As with all agricultural crops, grapes need a certain type of environment to thrive. Once a suitable location is found, the land is plowed and the vineyard is laid out by rows. Stakes are driven at measured distances, and a vine is planted next to each stake.

The next few years entail much work. The vine must be pruned and trained. Without this care the vine would waste nutrients and energy in growing cane and leaves instead of producing a good grape. With care the vine will produce grapes for a half a century or more.

The weather holds the key to a good or disastrous year. In the spring buds explode and tiny clusters of blooms appear on the plants. A bad frost at this point can kill the buds and destroy most of the year's crop. But with good weather the flowers will set fruit by the end of the spring.

The grape growers spend the summers weeding the vineyard and watching the grapes swell with water and sugar. By the end of the summer, the grapes are ready for harvesting.

As the grapes grow heavy on the vine, the winemaker must decide when to harvest. The grapes are monitored for just the right combination of sugar and acid. Generally, higher sugar content means higher alcohol content, which will add body to the wine.

The grapes may be harvested either by hand or machine. Handpicking grapes allows an expert eye to examine each bunch and to cull grapes that are damaged or too green. Mechanical harvesting, however, is quicker and cheaper and can get the grapes into the winemaking process before the sugar content changes.

The grapes are then destemmed, crushed, and pressed. At this point, however, different processes are used, depending on whether the final product will be a white, red, or rosé wine.

White Wine

White or red grapes can be used to make white wine. If red grapes are used, the skins are not allowed to stay in contact with the juice, so they will not lend their red color to it. Destemming, usually the first process in winemaking, may or may not be done when white wine is being made.

Stems contain tannin, which has a hard, astringent flavor. Some winemakers leave the stems on so the grapes may release the tannin flavor into the juice. If the wine is to be aged, the tannin will mellow with age and give more character to the wine. With or without stems, the grapes are then crushed. The crushing breaks the grape skins. Next, the grapes are pressed to release the juice. The juice, skins, and seeds are called *must* at this point.

The winemaker adds yeast to the must to begin the fermentation process, which converts the grape sugars into alcohol. Feeding on the fruit sugars, the yeast produces alcohol and carbon dioxide gas. The yeast becomes so active that the mixture froths as the carbon dioxide escapes from the tank. As more sugar is consumed, the process slows and the yeast slowly die (or are neutralized by the winemaker) and precipitate to the bottom of the tank.

The winemaker closely monitors the fermentation process by controlling the temperature and by other means. Without controls the liquid will continue to ferment until there is no sugar left. But if the must is too cool, most of the yeast will fall dormant and precipitate to the bottom of the tank, stopping fermentation. Temperature control is important for another reason as well. The fermenting must, if left alone, can hit temperatures as high as ninety degrees. The high temperature can cause the loss of some of the more delicate aromas and flavors.

With fermentation completed, the decision must be made whether to age the wine. Not all wine improves with aging. Some wine, especially whites, have their fruitiest flavor within months of fermentation. Other whites grow and improve from contact with oak barrels. Following fermentation the wines are filtered and transferred to either oak barrels or stainless-steel holding tanks to wait for bottling.

Fining also occurs at this time if needed. Traditionally, *fining* means adding colloidal agents, such as egg whites, to the wine. The egg whites slowly pass through the wine from the top of the vat to the bottom. Along the way they attract particles suspended in the wine. The egg whites take these particles with them as they go to the bottom, thereby clearing the wine. Another way to clear the wine involves sending it through a filter to remove the suspended particles. When needed, the wine is racked—that is, it is pumped from one barrel or tank into another, leaving behind the lees, the residue that settles to the bottom. Racking is done when the wine is cloudy due to grape proteins and yeasts that are suspended in the wine.

Finally the wine is bottled. The wine then "rests" in the bottle. Depending on the wine and the winery, some wine will age in the bottle for months or years, while others will be stored for only a few weeks before being shipped off to stores.

Red Wine

When a red wine is made, the grapes go from the crusher—seeds and skins alike—into the fermenting vat. The must ferments with the skins anywhere from a few days to several weeks. This is when the must takes on the pigment from the grape skins, giving the wine its color. The longer the skins remain in the juice, the darker the wine becomes. In all but a few cases, the coloring for wine comes from the grape skins, not from the grape itself.

Unlike the fragile whites, temperature is not as critical for reds. The must ferments, and the skins and seeds begin to float to the top of the container. The skins and seeds form a thick top layer, a cap, that the winemaker needs to break up and remix with the rest of the tank. This is done by pushing the cap down or by pumping the wine from the bottom of the tank to the top. After the desired fermentation has occurred, the skins and seeds, which give the mixture both color and tannin, are removed.

Unlike whites, most reds are aged in barrels to mellow their harsh flavor. Aging softens the taste of tannin, producing a mellower wine. The wooden barrels themselves also affect the flavor of the wine. Traditionally wine barrels are made of oak, but not all oak is created equal. The distinct flavor that each oak gives the wine can be too strong, depending on the type of oak used. This oak flavor, along with the tannin that is released from the oak wood itself, can overpower the bouquet of a wine. As the reds age, however, the strong tannin taste becomes less pronounced.

Rosé Wine

Rosés are made in much the same way as white wines. For rosés, however, the skins are left in contact with the juice for a short time during fermentation. This brief contact with the skins also gives additional flavor. Then the process for making rosé continues much like the process for fermenting a white wine.

Another method, though not used to produce the finest rosé, is to add red wine to a white wine to produce a rosé color.

Enjoying Wineries

As with anything else, your attitude and how you approach a winery can determine what you will get out of it. After visiting more than two hundred wineries, I have picked up several ideas and suggestions that I hope will make your visit more enjoyable.

The wine you taste in the South isn't the same as the wine in the West or the wine in the East. Yet each winery and winemaker has put a lot of work and effort into producing the product. Most tasting rooms provide a container to throw away the rest of your sample if you don't like the wine. There's no need to make faces or to say you think it's a terrible wine. Each person has different tastes.

Many wineries are small, family operations. A problem in the family may unexpectedly close the winery down for a day. And although the hours listed in this guide are as current as possible, some wineries change their hours without prior notice. Before driving long distances to visit, call and confirm hours of operation.

You may want to consider the time of the year. While some wineries want you to come and watch the harvest and crush, others will be too busy to show you around. It's a good idea to call first.

In busy areas, such as Napa Valley in California or the Finger Lakes in New York, you may want to visit the wineries early in the day before they get too busy.

Sometimes you'll be tempted to visit numerous wineries in one day, especially in places like Napa Valley, where the wineries are right next door to each other. After two wineries, though, the wines start to taste the same. If you're in a rush to visit the next winery on your list, you can miss truly enjoying the one you're at. You may want to schedule other events between wineries so each one will remain special. Also, before you know it, all those tastings can add up to quite a bit of wine. Take a walk through the vineyards, or eat your picnic lunch before getting back on the road. Don't drive drunk!

A few more tips for an enjoyable winery experience:

• Avoid wearing cologne, perfume, and after-shave. You don't want to drive hours to a winery and have your sample of wine ruined by the smell of Brut or Chanel No. 5.

• Some wine cellars are chilly; bring a sweater.

• Depending on the time of year, you may want to bring a blanket for a picnic—or your cross-country skis.

• Bring your questions. Most of the staff at wineries are well informed and love to share their expertise.

• Bring a notebook with you to write down the wines you really like and why. You can use it for reference the next time you visit your favorite wine store.

• Try something you never had before or something you wouldn't expect to like. Also, many wineries sell certain wines only at the tasting room. Try a sample of those. Ask the winemaker and the tasting room staff what their favorite wine is and give it a try. You may be in for a wonderful surprise.

• Don't let other people intimidate you about wine. If you like the wine, that makes it a good one.

• Although the winery may offer samples of all its wines, don't feel you have to drink them all.

• If the person pouring offers no direction on the order of sampling wine, you can follow the general rule of drinking drys before sweets and whites before reds.

• Finally, it's common courtesy to buy a bottle of wine.

The prices and rates listed in this guidebook were confirmed at press time. We recommend, however, that you call wineries before traveling to obtain current information.

The Wineries

STATE BY STATE

ALABAMA

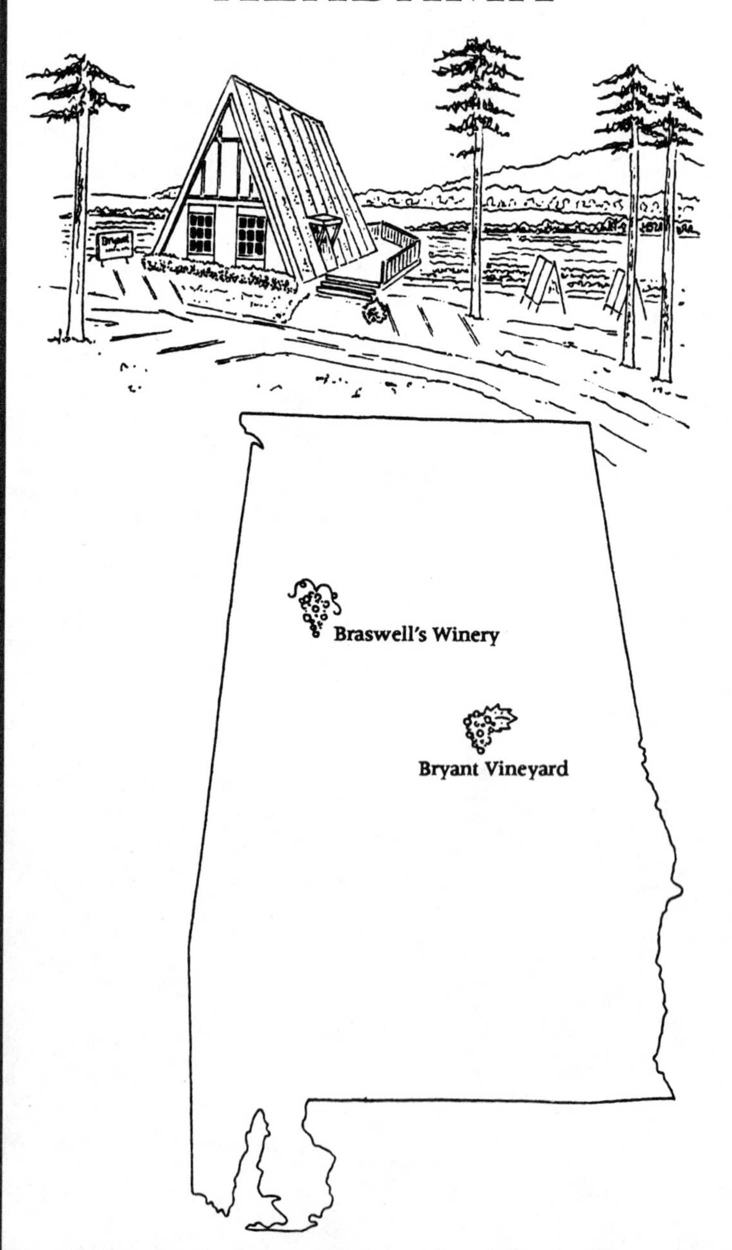

When explorers visited southern North America, Alabama, like other states, had an abundance of native grapes growing wild, primarily Muscadines. Now the most widely planted Muscadine is the variety Scuppernong. Through the years growers have experimented with native grapes and have developed Muscadine hybrids with such names as Magnolia, Noble, Carlos, and Dixie.

All native northern American grapes and European grapes grow in bunches. Not so with the Muscadines—they grow in clusters. Each grape of the cluster is large and round, rather like a marble. This makes a difference at harvest. Instead of picking bunches, harvesters spread a tarp beneath the trellises that hold the grape vines. Then, with something like a baseball bat, they hit the wire trellis, vibrating the ripe grapes so they fall off the vine and onto the tarp.

Another difference between the native American grapes and the vinifera grapes that are grown in California and other areas is the sugar content. Most mature native grapes do not have as high a sugar content when they ripen as that found in vinifera. Consequently, when the sugar is turned into alcohol, most sugar is lost. In order to achieve the desired alcohol content and to produce something besides a dry wine, winemakers sometimes sweeten the wine.

The hardy French-American hybrids are also grown in Alabama. The Bryant Vineyard, for example, experiments with Baco Noir and other hybrids. It also has experimental plantings of such vinifera varieties as Chardonnay, Pinot Noir, and Merlot.

Braswell's Winery

7556 Bankhead Highway
Dora, AL 35062; (205) 648–8335

If you can get juice out of it, more than likely the Braswells make a wine out of it. If not, well, they may just try it anyway. That's just the way they are.

Ruth and Wayne Braswell sell sixteen types of wines made from apples, blackberries, blueberries, cherries, elderberries, Muscadines, peaches, pears, persimmons, plums, strawberries, red grapes, white grapes, wild grapes, and native American varieties of Scuppernongs and Concords. Which is not to say that they make fruit wine because they can't make a good grape wine. In 1986 the president of the *Wine Industry News,* David Pursglove, mentioned the Braswell's Winery native grape table wine in his assessment of the best wines from wineries coast to coast.

Such recognition hasn't gone to Ruth's head. She's too busy keeping her customers happy. According to Ruth, a small wiry woman, "The plum wine is our number-one seller. The sweeter wines sell better."

To reach the winery you drive west of Birmingham through

rolling pine-covered hills. The winery building, a private club until 1982, has the feeling of a diner, right down to the stools and formica counter where visitors are served wine. Dog posters and pictures of dogs cut from magazines cover one wall of the building. Behind the counter a plastic sign illuminates the list of all the types of wine available, spelled out in little plastic letters. There's a mishmash of tables and chairs where you can sit if you don't want a stool at the counter. Or you can stroll through the display cases and tables filled with wine bottles, wine accessories, T-shirts, and stuffed animals.

If you want to see where it all begins, walk through the vineyards and fruit trees behind the tasting room. The inside of the processing room, also located out back, can be viewed by watching either or both of the Braswells' homemade videos. Wayne carries you, as he did the camera, through the pressing process, grapes on the vine, and plants at different stages of growth. Wayne was learning to use the camera when he made the videos, explains Ruth; that's why there are plenty of out-of-focus shots and shaky footage. It's only right. A slick, professionally made video doesn't belong here.

One part of the video shows how the cherry, in full blossom, became covered with snow in a freak storm that ruined their crop in 1987. "We wrote Montgomery for special permission to buy grapes in '87," says Ruth. Under Alabama law, at least 75 percent of the fruit used must be produced by the farmer. "I wish we didn't have to grow 75 percent—to cut down on my work." Except for certain times of the year, "like when a truckload comes in," says Ruth, it's just the two of them to take care of the 10,000-gallon winery.

The Braswells hand wash all their fruit and eliminate any bad pieces. Instead of filtering the wine, they rack it, transferring it from barrel to barrel, several times, leaving the sediment at the bottom. The wine clears as the sediment drops to the bottom. Wayne believes this takes the heavier visible particles out but leaves the flavor in, unlike commercial wineries that filter the wine many times.

The flavor can definitely be found at Braswell's Winery—from Wayne and Ruth down to the wines. The only drawback to visiting the winery is knowing that you've missed the Braswells' African gray parrot. He died in 1987. According to Ruth, "He got pancreas problems." The parrot filled the winery by squawking wine terms and hollering, "Drink Braswell's wine. It's good. Ya!"

DIRECTIONS: From Birmingham, drive west on Highway 78 for 22 miles. At milemarker 83, turn right on Bankhead Highway. After 1.5 miles, the winery will be on your left.

HOURS: Tours and tastings Thursday through Sunday from noon until 5 P.M. The winery is closed January, February, and March, and all major holidays.

EXTRAS: Gift shop sells wine accessories, T-shirts, wine, and beer- and winemaking supplies.

Bryant Vineyard

1454 Griffitt Bend Road
Talladega, AL 35160; (205) 268–2638

The Bryant Vineyard is a true down-home family affair. Three generations of Bryants work in the vineyard. The wine is made in the basement of their house, although they have plans to build a winery and tasting room "up the road." Until they do, if you take the time and trouble to find the vineyard, you'll be rewarded by a tour of the family's house (still under a little bit of construction) and shown how to make wine in a basement. And if you happen to stop by around lunchtime, you may be invited to stay and have a sandwich at the kitchen table.

But rather than taking advantage of their kind hospitality, bring a lunch, stop by and buy a bottle of wine, and then drive down and visit Logan Martin Lake. The lake has excellent swimming, waterskiing, picnicking, fishing, and boating. Or if you prefer trees to water, northwest of Talladega is a large section of the Talladega National Forest. The 364,428-acre forest offers long ridges with spectacular views of heavily wooded valleys filled with Southern pines and hardwoods. You can also swim and hike at the eastern side of the forest.

Leave time before your lunch and sightseeing for a tour of the Byrant operation. The Bryants, the owners, live near Tarrant, where Kelly, Jr. works as a fireman. If you happen to stop on a day when Kelly, Jr.'s working in Tarrant, then Kelly, the father of Kelly, Jr., and his wife, Kathleen, will be more than happy to help. Kelly and Kathleen live at the vineyard full time.

Grapes grew on the vineyard land long before it became the Bryant vineyard. "I've pulled down 50-foot Muscadine vines growing wild," says the older Kelly. In 1980 the Bryants changed from wild Muscadines to their own varieties of Muscadines. They also planted more than fifty varieties of vinifera, native American, and French-American hybrids, such as Villard and Seyval.

Harvest at the Bryants begins in September or October, depending on the variety. The family presses the grapes in a homemade yellow-pine press. The juice is then pumped into the basement through a tube, leaving the grape skins and stems outside. The Bryants use steel and oak barrels for fermenting and aging. They bottle the wine by hand.

The Golden Harvest, a sweet white, is "popular with the local crowd," according to Kelly, Sr. The Bryants also produce a Blush, Alabama Red, Country White, and Villard Blanc. "People in the area like sweet wine, but we like dry," says Kelly, Sr. So the Bryants make enough types to make everyone happy. They made the critics happy in 1990, when the winery's Dixie Blush won a top prize in the International Eastern Wine Competition.

The Bryants also plan on making a champagne in the future. You can take advantage of what they're producing now for $4.95 to $6.

DIRECTIONS: From Birmingham, take Interstate 20 east to exit 158. Follow Highway 231 south until you reach Highway 34. Continue on 34 until you reach Highway 43. Drive south on 43 for 2.6 miles until you see the signs COUNTRY CLUB ESTATES and REDSTONE. Turn right at the signs and drive 2 miles to the end of the road. Turn right; the winery is on the right after .7 mile. Look for the large BRYANT VINEYARD sign on the side and drive through the large farm gate. Continue on the gravel road until you see the vineyard on the right.

HOURS: Tours and tastings available by appointment, Monday through Saturday, 10 A.M. until 6 P.M. Closed major holidays.

EXTRAS: Winery sells wine.

ARIZONA

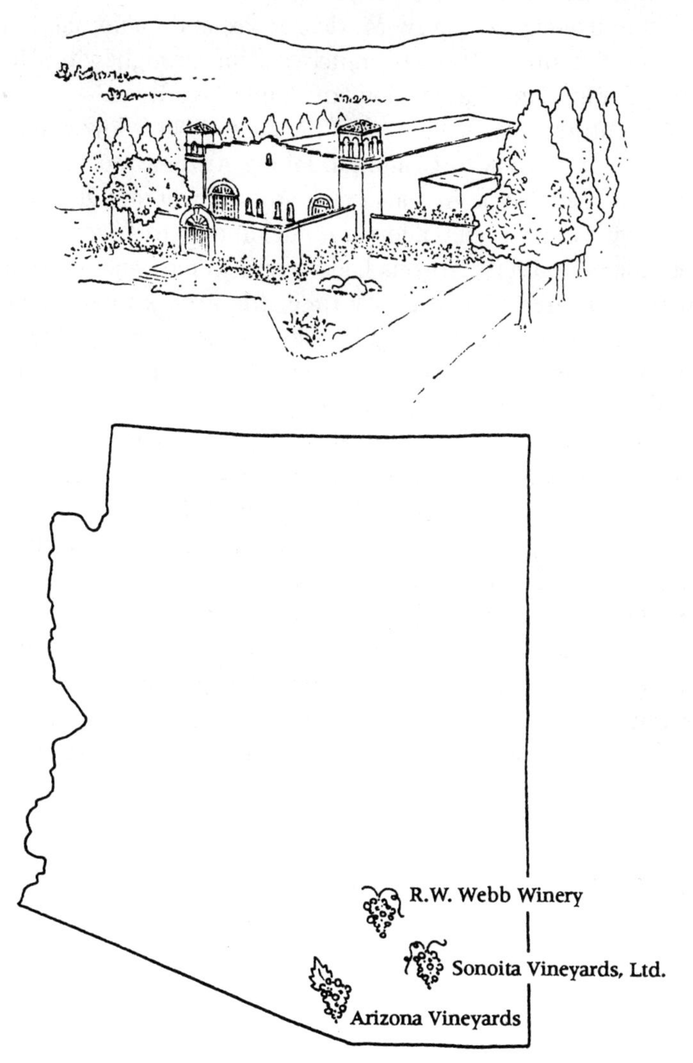

Spanish priests and explorers brought rootstock into Arizona, as they did to its neighbors, New Mexico, Texas, and California. In the seventeenth century Eusebio Francisco Kino brought vines from Spain and transplanted them along the Santa Cruz River.

Then, in the 1700s, Bautista De Anza and his fellow explorers traveled from Arizona to California, taking Arizona rootstock with them. As a result, some Arizona grape growers will proudly tell you that the wine industry in California is due, in part, to Arizona vines.

In the early 1900s, Arizona had more than ten wineries—mainly in the Southern Basin and Range, the area known as Four Corners. Phoenix, Chandler, and Nogales were producing wine, and many more areas were producing table grapes—putting Arizona in the top ten states in the nation producing grapes.

Prohibition shut down production, and as in many states, it has taken years for Arizona to recover. But in 1984 Arizona received official recognition for its grape growing when the U.S. Bureau of Alcohol, Tobacco, and Firearms named the Sonoita Basin, in the southeastern part of the state, between the Huachuca, Whetstone, and Santa Rita mountain ranges, a viticultural area.

Arizona has more than 200 acres of vineyards, and winery owners expect the figure to double soon. Wine growers in the region believe that it is only a matter of time until their product receives nationwide recognition and reaches the glory it knew almost 300 years ago.

Arizona Vineyards

2301 Patagonia Road
Nogales, AZ 85621 (602) 287–7972

"Welcome to the happiest place in the West. Every day is a happy day here," says Bob Bokk, as you walk into the winery. And it's true, not just hype. People who visit this winery have such a good time, they come back to work—to help Tino Ocheltree, the owner, harvest, ferment, and bottle.

"We have one lady from Germany that's been coming for years," says Tino. "We've had people stay for three or four days with us. If you come in and we're bottling, you're fair game," says Tino, smiling a devilish grin. What better way to learn about wines.

Because of the Arizona heat, the grape picking goes on from four-thirty until nine o'clock in the morning, and again from five until seven o'clock in the evening during June. No ultramodern equipment for this bunch. Tino says his winery is styled after a nineteenth-century rural European winery. He uses a handpress, and instead of closed stainless-steel fermentation tanks, he uses open concrete vats coated with beeswax. The vineyard gang has also been known to go to Portugal to harvest cork.

When Tino was visiting Bali, he bought a temple, crated it up, and shipped it back to Nogales. Now the carved winged horses, gargoyles, and gremlins decorate the winery. Says Bob Bokk, "Tino likes to collect." Because of the minimal lighting and collectibles everywhere, the winery appears to be a dark storage cellar of a world traveler who went on a spending frenzy.

Before the tour, or after the tour, or before *and* after the tour, you may have a free tasting. You belly up to a horseshoe-shaped bar and perch on a barstool. You can try any of the wine that's available—unless Tino decides to stop pouring it because the supply is running low and he wants to keep it for himself.

Or you can sample your favorite wine twice. You may want to keep track of what wine you are tasting, as you taste it. "If you lose your place, you have to start all over again," says Bob.

Even though there is absolutely no pressure to buy a bottle of wine, the Arizona Vineyards sells out. "We're a winery that's got a problem," says Tino. "We run out of wine." While other wineries would expand so they could sell more wine, Tino thinks that's silly. "Why do something bigger when you're happy?" He could raise the prices to keep down the buying, but he thinks that's even sillier. "Our prices haven't changed in years and years."

If you buy a case of wine, you'll take it home in a Budweiser beer box, or a box from a California winery. "We don't believe in spending our money on boxes," says Bob.

If you take your wine and wine tasting very seriously, you may want to consider skipping this winery. This is wine tasting with a sense of humor. According to Bob, the philosophy of the winery is to "make wine that looks good in a brown paper bag." The wine may look good in a paper bag, but it tastes great, too.

The first weekend in June the Arizona Vineyards throws an anniversary party. All the partygoers dress up in costumes and celebrate Padre Kino, the explorer who opened up missions in the area and brought Mission grapes to the state. During the party Tino releases three wines, and people retreat to the spring behind the winery, where they sit around bonfires and enjoy the winery's latest vintages. The group also carries a statue of Padre Kino back to the spring. "It's an all-nighter," says Bob.

Arizona Vineyards features such wines as the Workmen's Claret, Blanc de Blanc, White Burgundy, Mountain Rhine, Chardonnay, Zinfandel, Apache Red, Tino Tinto, and Cabernet Sauvignon, for $5 each. For a wine called Rattlesnake Red, Tino says they "even throw a rattlesnake in the vat."

But the people at Arizona Vineyards do more than just have a good time. Tino is serious about making a wine that he and others will enjoy drinking. The Arizona Vineyards' wines have done well when matched up against other Arizona wines. Not too concerned about awards, Tino hasn't entered wine competitions elsewhere.

DIRECTIONS: From Nogales, drive 4 miles east on Highway 82; winery is on the right.

HOURS: Tours and tastings 365 days a year, 8 A.M. until 5 P.M.

EXTRAS: Wine sold at the winery.

R.W. Webb Winery

13605 East Benson Highway, P.O. Box 130
Vail, AZ 85641; (602) 629–9911

If you catch Bill Morhous, the R.W. Webb marketing director, for a Sunday afternoon tour, he's liable to give the tour with a cup of coffee in his hand. And he will make sure you have a glass of wine, if you like, thus setting the informal tone of the thirty-minute tour. You'll receive a basic description of how wine is made as you wind your way through the press, fermentation tanks, and wine cases.

You'll hear descriptions of what the crusher does: "It replaces the feet"; and what happens with yeast in the fermentation process: "You could go down to the store and buy Fleishmann's [a bread yeast], but your wine would taste like bread." And Bill will tell you the wine has to pass a tough test before it's bottled: "We take off a gallon and seriously study it." This tour would be interesting even for family members who aren't particularly interested in wine.

A tour begins as soon as a group has gathered, or on the hour. While you wait for enough people to begin the tour, you'll be offered wine to taste. You'll also be given a wine list and pencil so you can check off the wines you prefer or want to remember. The winery produces a full line of reds and whites, from dry to sweet. Prices range from $4.35 for the Arizona Apple Wine to $12 for the 1986 Cabernet Sauvignon. They also sell wine accessories and T-shirts.

R.W. Webb, who displays his name prominently on each bottle, opened the first winery in Arizona since Prohibition, in 1980. He focuses on premium wines. In the past three years his wines have won thirteen medals, including a gold at the Les Amis Du Vine competition and four silvers at the Eastern International Wine Competition.

Bill will tell you that wine people in Arizona expect big things from their new industry—wine comparable with the great wines of Europe. "Everyone thinks of a desert. We don't try to grow grapes in the desert," Bill explains. "This is like the soil you'd find in the Burgundy area of France—*terra rossa.*"

DIRECTIONS: From Tucson, drive 20 miles on Interstate 10 east to the Vail exit. Turn left to cross back over I–10 and turn right onto the frontage road. After 1.4 miles the winery is on your left.

HOURS: Tours and tastings are offered Monday through Saturday from 10 A.M. until 5 P.M. and Sundays from noon to 5 P.M. The winery is closed Easter, Thanksgiving, and Christmas. Admission is $1, refundable with a wine purchase.

EXTRAS: Gift shop sells wine, wine accessories, T-shirts, and hats.

Sonoita Vineyards, Ltd.

P.O. Box 33
Elgin, AZ 85611; (602) 455–5893

"We produce a quality product, and we want people to set aside their biases and just try us," says Robin Dutt, the feisty and friendly manager of Sonoita Vineyards. "We are talking the best quality wines anywhere, but people don't believe you because we're in the desert."

And it's not only Robin who thinks Sonoita makes great wine. Its Pinot Noir and Cabernet Sauvignon have won major international awards. Barbara Ensrud, in *American Vineyards,* says, "Professor Dutt has already produced the most interesting and complex wines to come out of the Southwest."

Robin's father-in-law, Gordon Dutt, the winemaker, didn't set out with the single goal of making "interesting and complex wines." As Robin explains it, Gordon, who is a professor of agriculture at the University of Arizona, was looking for an alternative crop to anything else Arizona grows that doesn't take much water. He believed grapes could be an alternative crop for Arizona, and he was determined to prove it could be done.

In 1975 Gordon and A. Blake Brophy of the San Ignacio de

Babocomari Ranch began planting vinifera. After experiments with more than twenty varieties, the winery was founded in 1983.

Gordon's son, Richard, better known as "Rocky," began helping him on weekends in the vineyards. Then Robin joined in. "You learn everyday," she says about working at a winery. Although Gordon hasn't had much time to learn winemaking, Robin believes the winery has a bright future because Gordon "has a good nose and a good palate."

The winery sits atop a grassland hill that is surrounded by Mustang Mountain, Santa Rita Mountain, Huachuca Mountain, and the Canelo hills. "We terrace the land to catch water," says Robin. She will tell you other methods they use to grow grapes in Arizona on the fifteen- to twenty-minute tour of the winery. If you're interested, you can receive another fifteen- to twenty-minute tour of the vineyards as well. You will be exploring the vineyards in a truck, so wear appropriate clothes.

Unlike many vineyards that can't accommodate tours, Sonoita welcomes them. "For a large group we will use the tractor and put benches on the trailer. But first we'll let you taste all you want," says Robin.

The "tasting room," in the winery building, consists of a long slab of wood covered with a tablecloth and held up by oak barrels. Robin's sense of humor is not as dry as her wine. "We'll start at the dry end and work our way to the sweeter wines, and we'll all be happy." After she has poured the wines, she will encourage you to take a glass she's just poured. "Just grab and growl," says Robin to those who hesitate to pick up a glass.

You can "grab" Cochise County Colombard, Arizona Sunset (a blush), Arizona Dry Chenin Blanc, Sonora Blanca, Sonora Rossa, or a Sauvignon Blanc. All the Sonoita wines are available for tasting except its most expensive, the Cabernet Sauvignon. The wines are priced from $4.90 to $14. To taste the Cabernet you can buy a bottle at the winery, or attend the Cabernet Sauvignon release party held each year the Sunday before Thanksgiving. Sign up at the winery, and Robin will send you an invitation.

Don't expect the kind of Cabernet Sauvignon parties that are thrown in Europe. "We don't have a lot of pomp and circumstance, but we have a quality product," says Robin.

DIRECTIONS: From Tucson, drive east on Interstate 10 to the Sonoita-Patagonia exit, Highway 83. Continue on 83 for 25 miles to the junction of 83 and 82. Stay on 83 for 4.1 miles to the Elgin turnoff. Turn left and drive for 5 miles to Elgin and then follow the winery signs. After Elgin, the winery is 2.7 miles, bearing right after you pass through Elgin.

HOURS: Tours and tastings are available every day from 10 A.M. to 4 P.M. Closed major holidays.

EXTRAS: Wine is available at the winery.

ARKANSAS

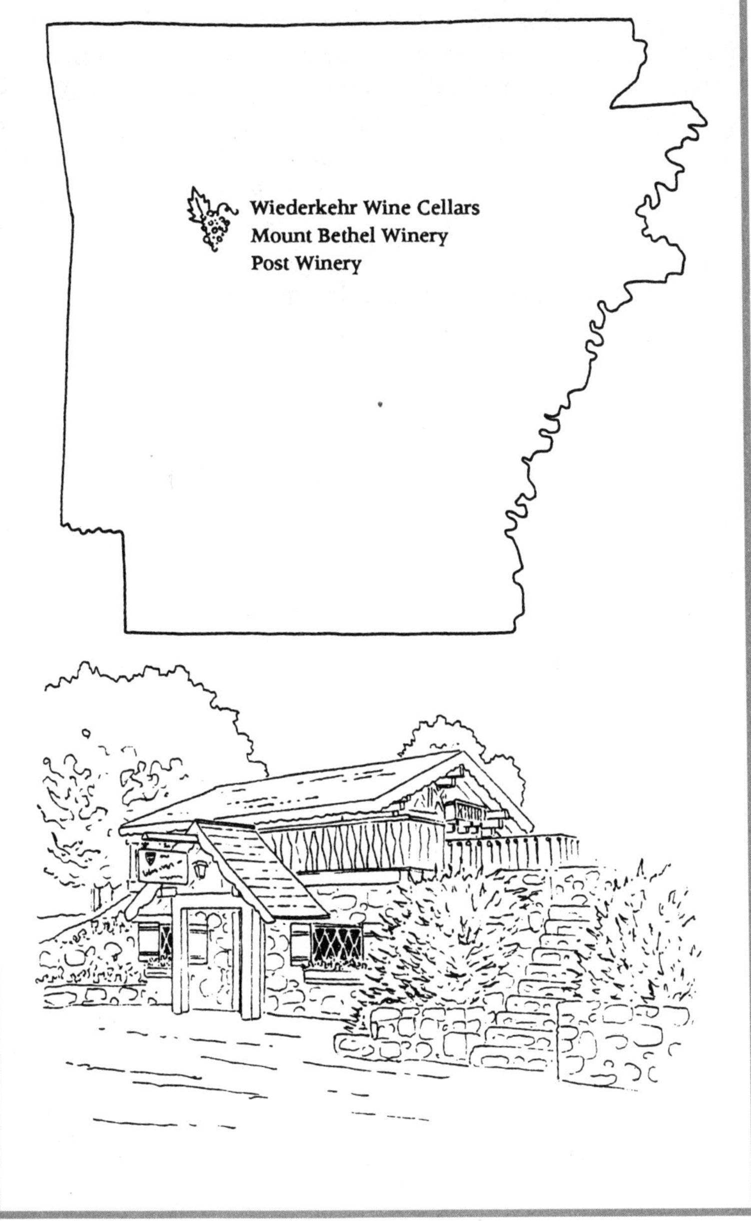

This is a community known for its wines," says Mount Bethel Winery's Peggy Post about the Altus area. It was the same way a hundred years ago. In Arkansas's northwest corner, German and Swiss immigrants brought their winemaking talents to the Arkansas River Valley and set out to recreate the vineyards they knew in their homelands. They settled in the Altus area in the 1880s. The wineries are still located in Altus and operated by descendants of those who planted vines. Little did the first settlers realize that one day the area would be producing almost a million gallons of wine a year.

Altus sits in a perfect location for growing grapes. To the north the Ozark mountains provide protection from cold winter winds, and to the south the Arkansas River and its valley help moderate spring temperatures and protect the vines from early frosts. Because of the area's history, location, soil, and climate, the federal government in 1984 designated Altus an official viticultural region.

The two major grape-growing families were, and are, the Wiederkehrs and the Posts. The Wiederkehrs have created the Wiederkehr Village, which hosts a wine cellar and a restaurant. Post families operate the Post Winery and Mount Bethel Winery. Each winery is as different as native American Concord and European vinifera Chardonnay, and you'll enjoy comparing the different approaches to making and selling wine.

The whole town of Altus celebrates its winemaking heritage the first Saturday and Sunday in July with the Altus Grape Festival. In the city park local winemakers and vineyard owners hold tastings. Also on hand are grape stomps, amateur wine- and jelly-making contests, waiter and waitress tray races, food, entertainment, and arts and crafts. "It's been very, very successful," says Peggy.

In addition to the wineries, the area offers paradise for those who enjoy the outdoors. Ozark Lake on the Arkansas River has some of the best fishing in the state. Bird-watchers can try to spot the more than seventy eagles that make the Arkansas River Valley their winter home. For the more active, the Mulberry River offers exciting white-water rafting. Just north of Altus is the 100,000-acre Ozark National Forest, with camping, hiking, fishing, and canoeing.

Mount Bethel Winery

U.S. Highway 64
Altus, AR 72821; (501) 468–2444

"Doesn't that smell nice? We get addicted to it," says Peggy Post as she enters the winery where grapes ferment in the fall and give off a wonderfully yeasty smell. Peggy and her husband, Eugene, who own the winery, come out to the building just to sit and enjoy the aroma. But they don't spend a lot of time sitting. "He'll spend half of October pumping [wine from] one tank to another and whooshing out the sediments," says Peggy as she explains how the winery operates.

Some of those tanks hold more than 7,000 gallons, and are made from California redwood. Although the tanks are large, Peggy doesn't see Mount Bethel as a large winery. "While other wineries are busy growing bigger, we are endeavoring to grow better, one bottle and one wine at a time," she says. "We've kept this down to family size."

The winery has always been family-sized. "This winery was my husband's grandparents', Joseph and Catherine Post, Swiss-German immigrants who came to this region in the 1880s to farm. With them they brought tradition and customs, one of which was winemaking." Joseph soon planted vineyards to make wine for the family's consumption. Then he sold wine to the neighbors, until Prohibition shut down the winery. It didn't shut it down completely, though. Peggy says they found a false front on a wooden wall that had about three feet behind it, apparently used during the "dry" years. "You couldn't tell there was a secret room back there," she says.

After Prohibition, Eugene's father began the Post Winery, down the road from where Joseph and Catherine began. When his father died in 1952, Eugene returned home from college after receiving a bachelor's degree in chemistry. He worked at the Post Winery for four years before breaking off and beginning Mount Bethel, at the Posts' original family home. "We renamed this Mount Bethel after the area where most of the vineyards are," says Peggy. "We have about twelve acres now."

The Posts of Mount Bethel use only grapes and fruits grown locally, mostly from their own vineyards, as well as several types that grow in the wild. Their wines include Seyval, Niagara, Catawba, Concord, Cynthiana, and rosé. The winery is also known for its rich dessert and fruit and berry wines, which include blackberry, blueberry, wild elderberry, peach, wild plum, and strawberry.

Although the wines are economically priced, Peggy says no shortcuts are taken when producing the wine. "My husband works with it like he's going to drink every drop of it himself," says Peggy about the use of pesticides and insecticides.

The historic stone wine cellar, which was built into the hillside by Joseph and Catherine Post around the turn of the century, is now used as the sales and tasting room. It has the cool and aromatic atmosphere unique to a wine cellar. They have built a room onto the original cellar where the wine is now bottled.

Over a glass of the rosé or a taste of Niagara, Peggy will tell you the history of winemaking in Arkansas, if you're interested. Or she can tell you in two sentences. "There used to be a lot of wineries in Arkansas, but one by one they've died off," says Peggy. "We've been here forever, and we enjoy it."

DIRECTIONS: From Fort Smith, drive east on Interstate 40 to exit 41. Drive south on Highway 186. At Highway 64 turn left; after .3 mile winery is on the left.

HOURS: Tours and tastings offered Monday through Saturday from 8 A.M. until 8 P.M. and Sunday from 9 A.M. until 5 P.M. Closed Christmas.

EXTRAS: Winery sells wine and wine accessories.

Post Winery

Route 1, Box 1
Altus, AR 72821; (501) 468–2741

"Post—It's not just for breakfast anymore," say the T-shirts referring to the cereal and the Posts' wine. They may joke on their shirts, but when it comes to wine, they're very serious, and have been for years. The Post family has made wine since 1880, and five generations have worked in the vineyards. Jacob Post, a Swiss-German, began by making wine for his family and later sold it to neighbors.

"I've got three brothers and a sister I work with, and dad," says Paul Post, vice-president of marketing for the winery, talking about siblings Andrew, Tina, Thomas, and Joseph. Mathew, Paul's father, and the rest of the Post family, have added to the native American varieties Jacob used, and they are trying to produce world-class wine with French-American hybrids and vinifera. Using only Arkansas grapes, the Post Winery makes more than twenty types of wine and produces 100,000 gallons annually.

"We're planting some acres in Cabernet Sauvignon. Our future is in the premium wines, so that's why the Cabernet," says Paul. He adds with pride, "We're making a great champagne with the Chardonnay." Most vineyard owners in the area thought vinifera, such as the Chardonnay, couldn't be grown in the area due to the humidity and cold. "We've had problems with mildew," says Paul, "but no winter problems."

The winery offers estate-bottled dry wines such as Cabernet Sauvignon, Chardonnay, Seyval Blanc, Cynthiana, and champagne. Its semidry wines are Vidal Blanc, Vignoles, Ives Noir, Ozark Mountain Blush, and Brut Champagne. There's also an array of sweet and semisweet: Muscadine Red, White, and Blush; Delaware; Catawba; Maidens Blush; Niagara; and Sherry.

Most wines sell in the $4 to $8 range. The sparkling wines are more expensive, from $12.50 to $15. The grapes come from the family's vineyard 3½ miles from the winery, atop St. Mary's Mountain. The wines from St. Mary's have won awards from the International Eastern Wine Competition, the *Dallas Morning News,* and the Midwest Wine Competition.

To enter the stone-and-wood tasting room, you pass under a vine-covered entrance, which is heavy with grapes in the fall. You're given a wine glass with the winery's name etched on the side of it. You'll be asked whether you're partial to sweet or dry. The pourers give suggestions and offer explanations of the various wines. You're offered cheese and crackers to munch between samples. The tasting room is beautiful and features grape and vineyard scenes in stained glass by artist Bill Driver.

The winery is as modern as anything you'd find in Napa Valley. The wood winery building behind the tasting room is filled with stainless-steel tanks. There are also several giant redwood tanks. "They've been in use since Prohibition," says Paul. "We prefer the stainless steel," he jokes. "They're easier to keep clean."

You walk to a second building to receive the description on how the Post Winery makes its champagne in the *méthode champ-*

enoise. On the way back to the tasting room, finishing your tour, you pass the bottling line.

DIRECTIONS: From Fort Smith, drive east on Interstate 40 to exit 41. Drive south on Highway 186; winery is on the left.

HOURS: Tastings available Monday through Saturday from 8 A.M. until 7 P.M. Open Sunday from noon until 4 P.M., April to October. Tours available Monday through Saturday from 11 A.M. until 4 P.M.

EXTRAS: Winery sells wine, wine accessories, grape juice, blueberry juice and jelly, Arkansas-made gourmet foods, picnic items, and local handicrafts. Picnic facilities are available.

Wiederkehr Wine Cellars

Wiederkehr Village
R.R. 1, Box 14
Altus, AR 72821; (501) 468–2611

It's not just a winery, it's an incorporated village. Wiederkehr Wine Cellars offers a 112-seat restaurant, a banquet area, retail wine shop, gift shop, and electrical hookups for recreational vehicles.

Driving up and down the hills between Interstate 40 and Altus, you'll pass farmland and cows. As you top St. Mary's Mountain, suddenly you're in Switzerland. As you get out of your car, gentle strains of Alpine music waft toward you, blown out of hidden speakers. You have arrived at the village.

All of the buildings are built of wood in Swiss Alpine style. Costumed maidens greet you and direct you to the building you're looking for. Perhaps the Vintage 1880 Shop, where wine tastings go on and imported cheese and sausage are sold, is the place for you. Those looking for a meal will want to head for the Weinkeller Restaurant, which serves Swiss cuisine and is listed on the National Register of Historic Places. Or if you're looking for a tour, step over to the Gift Shoppe, where the ten-minute tours begin.

While waiting for the tour, you're tempted by items for sale, such as wine accessories and Swiss cow bells. Don't worry about getting too involved in your shopping; a loudspeaker will warn you when a tour is about to begin.

Outside the Gift Shoppe, the tour guide starts with the history of Wiederkehr Wine Cellars. Swiss-German immigrant Johann Andreas Wiederkehr founded the winery in the 1880s. He felt at home in Altus, even though the foothills of the Ozarks aren't quite as high as his native Alps. He was followed by his son Herman J. B. Wiederkehr, and a string of Wiederkehrs. Al Wiederkehr is currently chairman of the board. "We're one of America's best-kept secrets," says Al. "We want to preserve the heritage of those that settled here."

The guide takes the tour group past the wine cellar Johann originally dug in 1880, which became a restaurant in 1964. Then it's through the banquet room, where family members' pictures hang from the walls. A walk to the cellars follows, with a description of the winemaking process. An interesting part of the tour is the opportunity to see 200-year-old German wine casks and casks of different shapes and sizes.

The tour includes a twelve-minute video showing winemaking from grape to glass, including grafting, pruning, harvesting, and all other phases. "On the video you can see all of it," says Al. "You'll leave with a complete view on how wine is made." And why Altus is special. "My grandparents knew what they were doing, picking this site."

Moving quickly to the tasting part of the tour, the crowd bellies up to a beautiful 30-foot tasting bar. From Salerno, Italy, the 140-year-old black walnut bar features gold-colored, hand-carved figurines. The group is given four wines to taste. You'll have to pay attention, though, because while you're tasting one wine, the tour guide may have already begun to describe the next. The tour ends with a question-and-answer period.

The winery offers a complete line of estate-bottled vinifera, Merlot, Pinot Noir, Johannisberg Riesling, Gewürztraminer, and Chardonnay. It also produces a semidry French Colombard; a medium-sweet, gold-medal-winning American Niagara; a lightly sweet Alpine Rosé; and sparkling wines. In total, it grows thirty varieties of grapes on approximately 300 acres of vineyards, makes more than forty different types of wine, and produces more than half a million gallons of wine each year.

If you'd like to eat before heading back to the highway, the Weinkeller offers Swiss fare. Breakfast, lunch, and dinner are served, and wine is available by the glass or bottle.

After your stay at the village, be sure to visit Wiederkehr's observation tower. North of the winery and 1.3 miles on your left, the 14-foot platform gives you a look at the vineyards and a 50-mile view of the rolling hills of the Arkansas River Valley. This is the view Johann had when he came to the area. You may understand why he decided to stay.

DIRECTIONS: From Fort Smith, drive east on Interstate 40 to exit 41. Drive south on Highway 186; winery is on the right.

HOURS: Tours and tastings offered from 9 A.M. until 4:30 P.M. daily and after 4:30 P.M. by appointment. Closed Good Friday, Easter, and Christmas.

EXTRAS: Gift shop sells wine and wine accessories.

CALIFORNIA

Simi Winery
Sterling Vineyards
Korbel
Champagne Cellars
Beringer Vineyards
Charles Krug Winery
Sebastiani Vineyards
Mirassou Vineyards
Thomas Kruse Winery
Mount Palomar Winery

Father Junípero Serra and other Franciscan missionaries brought winemaking to California as they moved northward from Mexico along the West Coast founding missions. Along with the missions came the Mission grapes. These grapes were used to make sacramental wine. Beginning in the early eighteenth century in Baja, the missions and Mission grapes moved north, and by 1769 they had reached San Diego. By 1823 twenty-one missions dotted California.

Also in the 1820s, near Los Angeles, Joseph Chapman started what has been called the first commercial vineyard in the state, with four thousand vines. In 1831 Jean Louis Vignes from Bordeaux, France, began planting something other than Mission grapes, the European vinifera.

Hungarian Count Agoston Haraszthy brought European vines to southern California. After a move to Sonoma in 1857, he planted and had success with his vinifera there. California's governor, John Downey, commissioned Haraszthy to bring back more European cuttings in order to expand grape growing in the state.

New areas opened up for vineyards. Napa and Sonoma valleys filled with grape vines. The wineries found a growing demand for their product during the gold rush. The acres in vines tripled during the 1860s and showed no signs of slowing in the 1870s. The next thirty years showed California wines winning international competitions.

The industry's future looked bright, but then dimmed suddenly, as phylloxera, a root louse, destroyed much of the industry in Sonoma, Napa, El Dorado, and Placer counties. Fortunately, a way to combat phylloxera was discovered; it consisted of grafting European varieties onto native American rootstock that was resistant to the louse. As the vineyards began to flourish once again, Prohibition took effect.

Prohibition closed many promising wineries. Very few vineyards remained open. Several did so by changing their varieties to table grapes; others made wine for medicinal and sacramental use; and some sold grapes to home winemakers.

The repeal of Prohibition in 1933 found the California wine

industry to be a ghost of its former glorious self. The 1940s and 1950s showed slow growth but an upswing began in the sixties—reflecting the growing demand for domestic wine and recognition that California wines could hold their own against French and German wines. New growing regions and new winemakers have given a boost to an industry that continues to grow.

Beringer Vineyards

2000 Main Street
P.O. Box 111
St. Helena, CA 94574; (707) 963–7115

When you take the tour at the Beringer Vineyards, the tour guide will tell you it's a historic tour—and history the winery has. Beringer's is the oldest continuously operating winery in the Napa Valley. Even Prohibition didn't interrupt as the winery stayed open making sacramental wine. In 1876 Jacob and Frederick Beringer began with 215 acres, and the vineyards of Beringer's have grown to be among the largest in the Napa Valley.

On the tour you'll visit the Beringer caves. Carved out of a stone hillside, the tunnels and caves provide the cool temperatures needed for aging. During the late 1800s, hundreds of Chinese laborers, working ten hours a day, six days a week, and for ten cents an hour, took six to ten years to pick out the stone tunnels. The tunnels go 70 feet below the surface and provide a constant 58-degree temperature.

Abutting the tunnels is the original winery, built with 2-foot-thick walls. The three-story building was built to use gravity to move the wine along. A road across the hillside above the tunnels brought grapes to the third floor. The juice then flowed to the second floor where fermentation tanks did their work. Then the wine would drop to the cool ground floor for oak-barrel aging and bottling.

The thirty-minute tour of the winery finishes in the Rhine House with the tasting. The Rhine House was built in 1883 by Frederick Beringer for $30,000. The seventeen-room mansion now holds the tasting rooms and gift shop. Built of California redwood and brick, the mansion is styled after Frederick and Jacob's family home in Mainz-on-the-Rhine, Germany. Listed on the National Register of Historic Places, the house is filled with the original stained glass, which has a different theme in each room. Don't miss the hunting scenes on the stairs to the Founders' Room, where you can try a two-ounce taste of Private Reserve and Limited Releases for $2 to $3 per glass. (It used to be Frederick's bedroom.)

In the tasting room, which was the dining room, you will be served three of the Beringer wines to taste. To taste more, visit the Founders' Room or buy a bottle of Chardonnay or Cabernet Sauvignon—two of the winery's most highly awarded wines. Or choose your favorite from a full selection of reds and whites, priced from $5.50 for the North Coast White Zinfandel to $19 for a Private Reserve Napa Valley Estate Chardonnay.

Don't expect to see any of the Beringers milling about. In 1971, with no more winemakers in the family, Beringers sold the winery to Wine World Estates, Inc.

Whether you are strolling through the hand-hewn caves or sipping a barrel-fermented Chardonnay, Beringer's is the perfect place to drink in the history and the glory of the Napa Valley.

DIRECTIONS: From Napa, drive north on Highway 29, the major artery of the valley. Pass through Yountville, Oakville, and Rutherford. On the north end of St. Helena is the winery, on the west side of the highway.

HOURS: Tours available daily May through October, 9:30 A.M. until 6 P.M., with the last tour at 5 P.M. November through April hours are 9:30 A.M. until 5 P.M., with the last tour at 4 P.M. Free tasting follows tour. Closed New Year's Day, Easter, Thanksgiving, and Christmas.

EXTRAS: Gift shop sells wine, wine accessories, wine books, T-shirts, and stemware.

Korbel Champagne Cellars

13250 River Road
Guerneville, CA 95446; (707) 887–2294

Think of champagne, and what comes to mind? Special occasions, celebrations, a tickled nose, and just plain fun. That's exactly what Korbel gives the visitor, too.

Korbel, with 500 acres of vineyards, produces 1.1 million cases of champagne each year—that's more than 2.5 million gallons—making Korbel the largest producer of champagne made in the *méthode champenoise* in the United States and fourth in the world.

West of Napa and Sonoma valleys, Korbel lies in the Russian River area between ridges of the Coast Range. As the Russian River winds its way through the valleys, it creates many microclimates excellently suited for grape growing. Wine has been produced in the area by people such as the Korbels for more than a hundred years. Most of the other wineries' grapes went into jug wines until the sixties, when more Americans "discovered" California wine and created a demand for more premium wines. Now there are numerous wineries in this area of Northern Sonoma, Lake, and Mendocino counties.

Even history is made fun here. Anton, Francis, and Josef Korbel didn't begin with the champagne cellars. They first tried their luck with a cigar-box factory and then a mill to cut the wood for the boxes and process redwoods. They later planted a fruit-tree orchard. Next they printed a paper with political cartoons called *The Wasp*, which was full of "stinging remarks." They also made bricks for the buildings on the land, because they could make them cheaper than buying them. Finally the brothers tried champagne. The first wine was produced in 1881 and the first champagne in 1882.

Beautifully manicured grounds with forest-green grass and roses offset the red-brick buildings with white trim. There is a garden you can tour during the summer months, with a white gazebo in an ocean of roses.

During Prohibition the champagne cellars stayed open by producing and storing wine for export. When repeal came, Korbel was ready to sell champagne immediately. According to the winery, the first case was shipped to President Franklin D. Roosevelt as a gift.

The last of the original Korbel brothers, Anton, died in 1925. His wife, Theresa, took the helm. After Theresa's death in 1938, Josef's daughter, Caroline, assumed the presidency. In 1954 Caroline and the rest of the Korbel family began looking for a buyer that would keep the Korbel family name and would be interested in maintaining Korbel's high standards of winemaking. They found a buyer in the Heck family. Adolf Heck ran the cellars at first, and now the winery is operated by Gary B. Heck.

The forty-five-minute tour begins with a walk by the Brandy Tower and an account of the Korbel brothers' exploits in Bohemia. The Korbel Brandy Tower is modeled after the prison in Bohemia where Francis Korbel was incarcerated after being arrested in a student demonstration in 1860. The story they tell at the winery is that Francis escaped by dressing up as a woman and just walking out of prison. The ivy-covered, red-brick Brandy Tower is complete with bars on the windows. The tower still stands, with help from steel bands that were placed around it after it was damaged in the 1906 San Francisco earthquake.

Then it's into the winery, where you're asked to fill out a card with your name, address, and your birthday (you're not asked for the year!). Korbel will send you a birthday card, and "remind you of the right way to celebrate," says the guide.

Old photos, plans for the cellars drawn on linen paper, and original copies of *The Wasp* are displayed. The guide explains some of the collection as you pass in front of the display case. Then an eight-minute slide show with beautifully colored artistic photos produced by the San Francisco Light Works describes Korbel's wines and the grapes that produce them.

Next you're led through the building and told how champagne is made. "All champagne is, is wine fermented twice," says the guide with a grin on her face. It's a simplistic description for such a complicated process. The tour shows machines that were used in the process; one, a French machine that slammed corks into bottles, was called the guillotine because it broke so many necks.

The tour is filled with fun champagne facts: The cork comes out of the bottle at 200 miles per hour; the bottle is filled with 100 pounds per square inch of pressure; and 1 bottle in 50 used to explode during the riddling process, whereas now only 1 in 3,200 usually breaks, and that is during the corking process.

You won't see the actual bottling machines at Korbel, but you can watch a short video that shows the production line. Most impressive is the old boardroom, completely furnished in redwood.

Don't forget to look up—the ceiling panels are exquisitely and intricately carved.

A fifteen-minute tasting follows the tour. The pourer explains each of the wines that are served and offers suggestions of what food would complement the wine. Usually three types of champagne are offered during the taste. If you are especially interested in another "flavor," however, feel free to ask. The pourers will try to accommodate you.

Korbel makes eight types of champagne. Korbel Brut is a medium dry wine and Korbel's most popular. It has a 1 percent dosage. (A dosage is a small amount of aged wine mixed with syrup; it determines the wine's sweetness.) The Extra Dry, which has a touch of sweetness with a crisp biting taste, has a 1.5 percent dosage. Both are made from a blend of champagne grapes.

The Blanc de Noirs, made from 100 percent Pinot Noir grapes, has a touch of color from the grapes. It is a dry champagne with a .5 percent dosage. Korbel Brut Rosé is produced from a blend of champagne grapes. Its pink color comes from the skins of Pinot Noir grapes in the blend. It is a slightly sweet wine with dosage of 1.5 percent.

The Natural, with a dosage of .5 percent, is also a drier champagne from a blend of grapes. There are also a dry Blanc de Blancs, made from 100 percent Chardonnay grapes (dosage .5 percent), and Korbel Sec, a medium-dry wine made from a blend of grapes (dosage 1.5 percent). Korbel also produces Rouge, which is a red champagne produced from a blend of Cabernet Sauvignon and Pinot Noir grapes. Currently this wine is available only at the winery.

The champagnes at Korbel have suggested retail prices ranging from $10 to $12, but there are always specials going on at the winery—some very good deals, too.

DIRECTIONS: From Santa Rosa, drive north on Highway 101 for 5 miles. Take the River Road exit and drive west for 13 miles; winery is on the right.

HOURS: Tasting room open daily May through September from 9 A.M. until 5 P.M., and October through April from 9 A.M. until 4:30

P.M. Tours available daily at 9:45 A.M. until 3:45 P.M., starting every forty-five minutes from May through September. From October through April tours are hourly from 10 A.M. until 3 P.M. Closed New Year's Day, Easter, Thanksgiving, and Christmas.

EXTRAS: Gift shop sells wine, wine accessories, T-shirts, picnic lunch items, and gourmet food. Picnic facilities.

Charles Krug Winery

P.O. Box 191
2800 St. Helena Highway
St. Helena, CA 94574; (707) 963–5057

Although Beringer's is the oldest *continuously* operating winery, Charles Krug Winery is the oldest operating winery in Napa. The wooden cider press that Krug used to make his first batch of wine is still there. Krug fled from Prussia, where his political leanings weren't appreciated, and came to America in 1849. After meeting Agoston Haraszthy, a leader in bringing European varieties to California, and seeing the climate in Napa, Krug set out to make wine from grapes grown in that area.

He moved to Napa in 1860, planted vines on land that he

received as part of his wife's dowry, and built his wine cellar in the area now known as St. Helena. He founded his winery in 1861. As the first to plant grapes in the Napa Valley, Krug was a true pioneer.

By 1890 the Napa Valley wineries were second only to Sonoma Valley. Krug shared his expertise with other pioneers in the area such as the Beringer brothers.

The winery faltered when Krug died in 1892. It went to his daughters and then his nephew Bismark Bruck. Prohibition struck another blow, and for the most part the winery shut down during this time, although some grape juice was made for home winemakers.

The winery continued in limbo until 1943, when the Cesare Mondavi family bought it. Cesare had his first experience with grapes as a wholesaler; later he expanded into winemaking. His son Peter joined the business by running the Sunny St. Helena Winery. Soon the family was ready to expand and bought the Krug Winery. Besides the Krug name, only the carriage house of 1881 and the old cellar of 1874 were still in usable condition.

The Mondavis' first vintage under the Krug label appeared in 1944. Before Cesare Mondavi died in 1959, he saw the Krug label start to recover its former glory.

Today Peter's son Marc follows the Mondavi and Krug tradition. Marc oversees the vineyards and winemaking. The family owns 1,200 acres of Napa Valley vineyards.

Krug is proud of its reputation of catering to wine enthusiasts and visitors. It offers two options for the drop-in visitor. The first is a tasting in which you may taste five of Krug's wines, some of which are older, private reserve Cabernets and other wines not usually found in stores. A member of Krug's staff pours and explains each wine as you taste at your own pace. You are also given a Charles Krug glass, which entitles you to future complimentary tastings. The special tasting costs $3 and is available from 10:30 A.M. until 4:30 P.M. in the Krug visitors center. This tasting is complimentary on Wednesdays.

The second option is a tour and tasting together. From vineyard to bottling room to tasting, the hour-long tour takes you through the history and production process of Krug. You walk among the vines, visit the buildings on the sprawling grounds, and strain your neck as you try to see the top of the towering 30,000-gallon redwood tanks.

You visit the historic 1874 winery building with 30-inch-thick walls. Before refrigeration the doors and windows of the winery were left open at night to let the cool air in, thus keeping the building at fifty degrees. Next you'll move on through other parts of the production process. As you wind your way through Krug's 6,200 French, German, and Yugoslavian oak barrels, a tour guide explains how Cabernet Sauvignons are made.

After the production tour you receive a guided tasting. Your tour guide will pour several wines and talk about the wines' characteristics. An extra bonus—the guide also explains how to evaluate a wine (what to look for and how to taste it) and suggests what wine might complement specific food.

You might receive a taste of the Grey Riesling from Napa Valley grapes. Krug calls it a "fruity, soft, off-dry wine perfect for sipping and luncheons"; it sells for $6.50. The popular White Zinfandel, made from a red grape, is known as a blush, slightly sweet. It can be enjoyed alone or with food and costs $5.50. The Cabernet Sauvignon, a consistent award-winner for Krug, sells for $9.

Other Krug wines include Chardonnay, Sauvignon Blanc, Fumé Blanc, Gewürztraminer, Muscat Canelli, Gamay Beaujolais, Pinot Noir, and Merlot. Prices range from $5.50 to $25 (prices vary for the special reserve wines). For the price, the quality, and the afternoon of touring and tasting, you can't pick a better winery in Napa to visit.

DIRECTIONS: In the Napa Valley, drive north on Highway 29, 1 mile north of downtown St. Helena; winery is located on the east side.

HOURS: Tours and tastings, for $1, offered daily, except Wednesdays, at 11:30 A.M., 1:30 P.M., and 3:30 P.M. Tastings only, for $3, available anytime from 10 A.M. to 4:30 P.M. Wednesday tastings complimentary. Retail room open from 10 A.M. to 5 P.M. daily. Winery closed New Year's Day, Easter, Thanksgiving, Christmas Eve, and Christmas.

EXTRAS: Gift shop sells wine and wine accessories.

Thomas Kruse Winery

4390 Hecker Pass Road
Gilroy, CA 95020; (408) 842–7016

Thomas Kruse says his winery is "the most famous unknown winery in the world. Not a day goes by that I don't hear 'I never heard of you before,'" says Kruse. Visitors may get the impression that he would prefer to keep it that way.

If he doesn't like the looks of you or if the wine-tasting room is getting too crowded, he won't let you in. If he does, you are in for a treat—a visit to a winery that closes for Super Bowl Sunday, which Kruse considers a major holiday. And you will meet the man who writes his own labels, such as this piece of eccentricity from the Johannisberg Riesling:

> *So You Want to Start a Small Winery. Assuming you have enough money and a little knowledge, allow me to supply what I believe are the essential elements for success in the wine business. 1. Put the winery on a hill and difficult to get to. 2. Visits should be by appointment only. 3. The winemaker should be aloof, haughty, arrogant and never be observed exhibiting unrestrained mirth. 4. The wines should be tannic, acidic, overoaked and*

unpleasant. Grotesque caricatures of the variety. 5. Last and perhaps most important price the wine obscenely—somewhere around a dollar a sip. Wait a minute! I think I went wrong somewhere!

And not all of his labels deal strictly with wine. This bit of wisdom may be found on his label for Zinfandel:

On Pasta and Life . . . I'm not sure when I first became aware that there are two kinds of people in this world—those who twirl their spaghetti and those who cut it. I've even developed definite opinions about this. First of all, twirlers are food people who love to eat. Cutters simply eat to refuel. Cutters only eat spaghetti, often breaking it before putting it in the pot and rinsing it after it's drained. Twirlers eat spaghetti and spaghettini, linguini, fedelini, vermicelli and fettucine. Breaking and rinsing are taboo.

Kruse received his license in 1971. His winery is a small one, producing 5,000 gallons a year. Kruse has two blends, a Gilroy white and a Gilroy red, and the varietals Zinfandel, Petite Sirah, Chardonnay, and Cabernet Sauvignon, available for $4.50 to $10. He also produces a sparkling wine made in the traditional French *méthode champenoise.* If he's busy with the sparkling wine, you may watch him work his magic on it.

Kruse will offer people a tour of the facilities "if I like them and they seem interested." Ask him to show you the drop weight corker that's more than one hundred years old. Tours are informal and available daily, except holidays, from noon to 5 P.M. If you want a tour and a full one-on-one dose of Kruse's wit, visit the winery on a weekday. The winery gets busy on weekends, so Kruse may get stuck in the tasting room and not be able to offer you a tour.

DIRECTIONS: From Highway 101 at Gilroy, drive 5 miles west on Highway 152; winery is on the left.

HOURS: Open daily from noon until 5 P.M., except major holidays and Super Bowl Sunday.

EXTRAS: Wine sold at the winery.

Mirassou Vineyards

3000 Aborn Road
San Jose, CA 95135; (408) 274–4000

On one of his trips from France to the United States, Pierre Pellier, the patriarch of the Mirassou family, was in danger of losing his grapevine cuttings. The ship was running low on water, and he needed to find a way to keep his cuttings moist. Using the resourcefulness that has kept the family business alive since 1854, Pierre bought the ship's entire supply of potatoes. He slit the potato ends and stuck the cuttings inside the potatoes for moisture.

Pellier came to California in the 1850s "looking for those gold-colored rocks we were growing here," says the tour guide at Mirassou. He soon found gold in the form of European vines. According to the winery's history, Pellier was the first to import the European varieties such as Pinot Noir and French Colombard.

The second generation in "America's oldest winemaking family," as the Mirassous call themselves, came when Pierre's oldest daughter, Henrietta, married Pierre Huste Mirassou in 1881. He died at an early age, thirty-three, but not before he left three sons to carry on the family business.

When the plant louse phylloxera devastated the California

vineyards, Mirassou was not exempt. But neither that nor Prohibition could close the winery. Henrietta and her sons pulled up the ravaged vines and replanted with native American rootstock. During Prohibition grapes from the vineyard were sold and shipped east. This gave the Mirassous a jump on other vineyards when Prohibition was repealed. Other vineyards had pulled up their vines and needed years until the new vines could produce a grape suitable for premium wine.

Henrietta's oldest son, Peter, brought his sons, Norbert and Edmund, into the business. They spent the next few years after Prohibition building it up. Edmund's sons, Peter, Jim, and Daniel, are the fifth generation at Mirassou. It was with this generation, in 1966, that the winery began selling wine under its name instead of marketing the wine in bulk to other wineries to sell.

This history is just one section of the four-part tour you'll receive at the winery. The tour also covers wine production, champagne production, and a brief explanation on how to taste and judge wine. The tour lasts anywhere from a half hour to an hour, depending on the interest of the group. "You can get a feel for what they want," says Frank Roberts, one of the guides at Mirassou. And the guides will go out of their way to make you and your group happy.

The tasting starts out with champagne. Mirassou makes four types of the bubbly—Brut, Blanc de Noirs, Au Naturel, and Brut Reserve. The first two sell for $13, and the Au Naturel and the Reserve for $15. It also makes a full range of award-winning wines, such as Chardonnay, Cabernet Sauvignon, Pinot Noir, and Johannisberg Riesling. Prices range from $5.50 for a varietal Sauvignon Blanc or Zinfandel to $13 for the Reserve Pinot Noir.

To showcase the wine Mirassou holds a series of events throughout the year, such as wine education classes, candlelight dinners with classical music, brunches, sunset dinners, and cooking classes. For a complete list of the more than forty annual events, contact the hospitality department at the winery. Perhaps when you visit the area you may take advantage of one of the events and make your visit even more enjoyable.

Also, you may want to include a side trip to the Mirassou

Champagne Cellars, approximately 20 miles away. In 1989 the cellars opened at the historic Novitiate Winery in Los Gatos to give Mirassou more space for champagne production and storage. The cellars are housed in a concrete winery built in 1893 by the Sacred Heart Novitiate Seminary. A beautiful example of the use of gravity in winemaking, the top floor of the winery was used to crush the grapes and then the juice flowed to the second floor for fermentation. The bottom floor, or cellar, was used to store and age the wine. The Jesuits ceased wine production in 1985, opening the way for Mirassou to lease the building.

The Champagne Cellars offers tastings daily from noon until 5 P.M. From Highway 17 exit at East Los Gatos. Turn right onto Los Gatos Boulevard, which becomes Main Street. Turn left on College Avenue and continue .5 mile to the cellars, 300 College Avenue (408–395–3790).

DIRECTIONS: From San Jose, take Highway 101 south to the Capitol Expressway exit. Head east until the second light, which is Aborn Road. Turn right on Aborn; vineyards are on the right after approximately 2 miles.

HOURS: Tasting available Monday through Saturday from 10 A.M. until 5 P.M. and Sundays from noon until 4 P.M. Tours available Monday through Saturday at 10:30 A.M., noon, 2 P.M., and 3:30 P.M., and Sundays at 12:30 P.M. and 2:30 P.M. Closed Easter, Thanksgiving, and Christmas.

EXTRAS: Gift shop sells wine and wine accessories. Picnic facilities.

Mount Palomar Winery

33820 Rancho California Road
Temecula, CA 92591; (714) 676–5047

"Our way of greeting the public, including the enthusiastic and informative tours, began with my father, John Poole," says Peter Poole, the general manager at Mount Palomar. "For many years he conducted legendary tours of the facility, and although he is now semiretired, his style still sets the tone. We don't believe in giving the 'canned' or highly commercial tours that one often receives elsewhere."

John Poole first planted vines in 1969 to create Long Valley Vineyards. He produced the first Mount Palomar wines in 1975, some of the first wine in the Temecula area.

His belief that wine should be enjoyed and not shrouded in snobbery and mysticism is seen and heard while touring the winery. The forty-five-minute tour begins in front of the tasting room, and you will see the whole operation, from smelling the yeast that is used to an explanation on how sherry is made. The tour takes you outside where the sherry is heated by the sun. You'll learn that sherry needs to be oxidized, a process that can take up to three years. Heat aids in the oxidation process.

"We get complimented on our tours," says Peter. "We update them [the tour guides] all the time if we are doing some things in the winery differently. We think of tours in terms of educational experience, not a sales pitch."

You may taste the wine before or after your tour. Tastings cost $2, but you may keep the wineglass with the Mount Palomar logo. The winery asks that you limit your tasting to four wines, and there's a 50-cent charge to try the 1986 Johannisberg Riesling Select Late Harvest. For $5 to $12 you can purchase award-winning estate-bottled wines such as the 1990 Chardonnay, 1990 White Cabernet, 1987 Cabernet Sauvignon, or 1989 Sauvignon Blanc.

Mount Palomar has a large picnic area. You can walk up a trail to one picnic area that affords a wonderful view of some of the winery's hundred acres of vine-covered rolling hills. In case you forgot to bring the makings for a picnic, the winery offers a full deli with gourmet salads, sandwiches, cheeses, and chilled wine. After learning a lot about winemaking and taste-testing your knowledge, you deserve a picnic.

DIRECTIONS: From San Diego, drive north on Interstate 15. Exit on Rancho California Road and drive east. After 5 miles look for winery sign on the left.

HOURS: Tastings available for $2 seven days a week from 9 A.M. until 5 P.M. Tours are held weekdays at 1:30 P.M. and 3:30 P.M. and weekends at 11 A.M., 1:30 P.M., and 3:30 P.M. Winery closed New Year's Day, Thanksgiving, and Christmas.

EXTRAS: Gift shop sells wine, wine accessories, and deli foods. Picnic facilities.

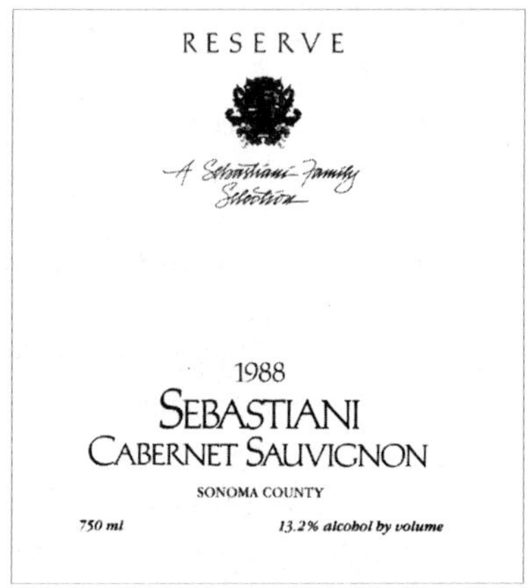

Sebastiani Vineyards

P.O. Box AA
389 Fourth Street East
Sonoma, CA 95476; (707) 938–5532

When the United States was less than fifty years old, vineyards were being established by Franciscan missionaries on land that is now known as Sebastiani Vineyards. In 1825 the friars at the Sonoma Mission established the first vineyard in Sonoma Valley and made sacramental wine. Sonoma is now perhaps almost as well known as its neighboring valley to the east, Napa. With its hot sunny days and cool ocean air, Sonoma provides the best weather conditions for varieties such as Chardonnay, Cabernet Sauvignon, and Pinot Noir.

After secularization in 1835 General Mariano Vallejo, commandant of the area, continued to produce wine. Samuele Sebastiani and his wife, Elvira, bought the property and produced their first wine in 1904. Today, with some of the original mission vineyard still planted in grapes, the Sebastiani vineyards are not only some of the oldest in the area but are one of the largest in California. Sebastiani Vineyards and Sylvia Sebastiani, the current owner, try to maintain quantity and quality. It appears to be working, judging by the number of awards the wines have won.

You can sample the award winners in the tasting room, located in part of the winery's original stone structure. The pourers suggest "you only try four or five wines, since some people have trouble differentiating after that"—one of the nicest ways ever heard to limit visitors' consumption.

If you take the tour before you taste, at the end of the tour the guide will give you a list of wines and explain each wine available for tasting. This will help you find a wine you'll most enjoy. Another benefit of taking the tour first is that the guide will explain any special wines that are available for tasting that day. But don't expect to find a bottle of 1941 Casa de Sonoma, a $100 Cabernet Sauvignon, open for tasting.

Sebastiani offers a full line of whites, reds, and sparkling wines you can taste. Prices range from $5 for Pinot Noir Blanc, Chenin Blanc, Gamay Beaujolais, Pinot Noir, and Zinfandel, to $25 for a 1982 Cabernet Sauvignon. The winery also sells sweet dessert wines: Muscat Canelli, Dolcino Rosso, Late Harvest Johannisberg Riesling, and Amore Cream Sherry. The last three dessert wines are available only at the tasting room.

In the tasting room and throughout the winery, look for the work of the "human termite." Earle Brown, a wood-carver, came to work at Sebastiani when he was sixty-six years old. For the next seventeen years he carved like crazy, leaving almost no piece of wood in the winery untouched. Most of the carvings are of birds—a passion of one of the second-generation Sebastianis. The redwood tanks have carvings of canvasbacks, pintails, and whistler swans.

Brown also carved an intricate twelve-paneled "Vintner's Calendar," showing what work happens each month at the winery: pruning done in January, spraying the vines in July, fermenting the grapes in October, and the best part of all, tasting the wine in December. You can see the sculpture, along with a glass case with Brown's tools, while you wait for the tour to begin. In this waiting room is also the original winepress that Samuele used in 1904.

The twenty-five-minute tour takes you from the original stone building throughout the newer winery buildings. Redwood tanks, built in the 1930s, hold from 4,000 gallons up to an impressive 59,666 gallons. The tanks' redwood staves, held together only with metal bands, are incredible to stand by. You have to lean back and crane your neck in order to see the tops of these giants. The rest of the tour explains how to produce wine, although sometimes it's hard to concentrate, thinking of how many good years a wine aficionado could have with just one redwood tank.

DIRECTIONS: Entering Sonoma from the south on Highway 12, turn right on East Napa Street, then left onto First Street, then right on East Spain. The winery is on the northeast corner of East Spain and Fourth Street East.

HOURS: The tasting room is open daily from 10 A.M. until 5 P.M. Tours are offered from 10:20 A.M. until 4:20 P.M. at twenty-minute intervals. Winery closed New Year's Day, Good Friday, Easter, Thanksgiving, and Christmas.

EXTRAS: Wine and wine accessories are sold at the winery. Picnic facilities.

Simi Winery

P.O. Box 698
16275 Healdsburg Avenue
Healdsburg, CA 95448; (707) 433–6981

Like a fine old Cabernet Sauvignon or a crisp young Riesling, both the old and the new have their place and their admirers. Such is Simi Winery, where old traditions coexist with new technology and result in fine wines and an enjoyable afternoon of touring. As for the old, Simi wines started being produced more than a hundred years ago; as for the new, the winery began a three-year, $5-million renovation in the late seventies and early eighties.

Simi Winery was established in 1881, when two Italian immigrants, Guiseppe and Pietro Simi, bought a winery in downtown Healdsburg. Pietro ran the brothers' produce business in San Francisco while Guiseppe took charge of the winery. By 1882 Simi had grown to the third-largest winery in Healdsburg, producing 100,000 gallons.

As the demand for their wine increased, the brothers decided to expand. They bought land north of Healdsburg, calling it Montepulciano Winery in honor of the winemaking region in Italy where they were born. As the wineries were in the process of expanding their capacity to 400,000 gallons, both brothers died, Pietro in July 1904, and Guiseppe one month later.

Pietro's family took control of the San Francisco business while Guiseppe's teenage daughter, Isabelle, assumed the huge undertaking of running the winery. The original Simi Winery closed after the earthquake of 1906, and Isabelle concentrated her efforts at Montepulciano. She ran the winery basically alone until her marriage in 1908 to a bank teller, Fred Haigh. He eventually quit his work at the bank to help his wife.

Isabelle and Fred didn't have much time to run the business before Prohibition struck. Although the winery was able to stay open by making sacramental wines, it did not earn enough money to pay the bills, and many acres of vineyards were lost. But when repeal came in 1933, Simi was ready. Isabelle and Fred had decided not to sell the half-million gallons of wine that Simi had on hand when Prohibition went into effect. So Simi was in a position to sell, while others had only very young wine or wine of poor quality. With repeal came a name change as well; Isabelle began selling all the wines under the Simi name.

The late thirties and early forties brought Simi recognition as a producer of fine wines. Fred became ill in the late forties and died in 1954. Once again the winery fell on the shoulders of Isabelle. The winery continued but it became increasingly difficult for Isabelle, especially when her only child, Vivien, died in 1968.

Russell Green, a fellow vineyard owner, bought Simi in 1970. The next few years brought several different owners to Simi: Scottish & Newcastle in 1974; Schieffelin & Company in 1976; and Moet-Hennessy in 1981, when it purchased Schieffelin. Throughout the ownership changes, Isabelle could be found in the tasting room, talking to visitors. Even after selling the winery, she remained a part of it until her death in 1981.

Schieffelin did the most to change the face of Simi. Its $5-million renovation included adding a 14,000-square-foot fermentation cellar, fifty-six stainless-steel tanks that can hold from 500 to 12,000 gallons, and four 4,500-gallon stainless-steel tanks. These changes helped Simi become a large producer of quality wines that are sold internationally.

Schieffelin also renovated the original stone cellar. Completed in 1890, the hand-hewn stone was first cemented together by Chinese workers. Italian laborers completed the addition in 1904. The two groups of workers had completely different ideas of how a building should be constructed, and the difference is quite obvious as you look at it from left to right. One side has the mortar between each rock placed neatly, while the other appears as if the workers simply flung the mortar at the building stones.

Schieffelin realized that the cellar's 3-foot-thick walls still provided an ideal way to maintain the proper wine storage temperature, so it had the walls stabilized and added support beams. The four-floored cellar holds more than 7,000 sixty-gallon oak barrels. Its combination of wooden floors, wooden barrels, and stone walls are a beautiful sight.

The cellar is where the forty-five-minute tour begins. Outside the building your guide tells of the winery's history and points out which side of the cellar was built by which group of workers. You also see the interior later in the tour. Between the two tours you are treated to a complete explanation of the winemaking process. You walk past "The Sugar Shack," where the grapes are tested for their sugar content, and into the production building, where the latest technology is used.

The end of the tour takes you past the old retail tasting room. In 1936 Isabelle had an enormous redwood champagne tank rolled out of the winery and converted to a tasting room. She had shingles added to the outside for protection, creating a unique structure. The new retail tasting room is an octagon-shaped building of stone and wood.

In the tasting room you may sample some of Simi's wines, such as Cabernet Sauvignon, Chardonnay, Sauvignon Blanc, Chenin Blanc, and Rosé of Cabernet. The wide range of Simi's wines may be tasted free of charge, except for several reserves. A taste of the reserves costs two or three dollars, but don't think you won't taste fine wine without paying for it. Simi's Cabernet Sauvignon and the consistently award-winning double-fermented Chardonnays will make most wine drinkers very happy.

DIRECTIONS: From Santa Rosa, drive north on Highway 101 to the third Healdsburg exit, Dry Creek Road. Head east to the second traffic light. Turn left (north) on Healdsburg Avenue. The winery is one mile north, on the left.

HOURS: Tastings available from 10 A.M. until 4:30 P.M. daily. Tours held daily at 11 A.M., 1 P.M., and 3 P.M. Closed New Year's Day, Easter, July 4, Thanksgiving, and Christmas.

EXTRAS: Wine and wine accessories sold at the gift shop. Picnic facilities.

Sterling Vineyards

P.O. Box 365
1111 Dunaweal Lane
Calistoga, CA 94515; (707) 942–3456 or (800) 999–3801

You've fought the stop-and-go traffic on Highway 29. You've been pushed and shoved as people try to get to the tasting bar before you. Need a break from the mad rush of Napa Valley? Sterling's the place.

Touch the treetops, soar with the birds, have Napa Valley laid out at your feet, all from a quiet glass cage—that is, Sterling's aerial tram ride. From the winery parking lot, the $5 ride glides you past Sterling's pond and fountain, complete with lily pads, up and away for an incredible view of vineyards.

The stark, white-stucco, Mediterranean-style winery and visitors' center greet you as you arrive at the top of the 300-foot-tall knoll. You start the self-guided tour on a balcony that provides another panoramic view of the valley. This tour consists of seventeen panels that explain the winemaking process. The first panel describes the soil's importance in the quality of wine.

"The dissolved minerals," reads the panel, "absorbed into the grape and alchemized in the winemaking process, provide the compound flavors that distinguish outstanding wines. Thus some of the world's finest vineyards stand on seemingly barren soil or cling to precarious slopes."

The next panel deals with climate: "However, seasonal differences in rainfall, in heat spells, and in periods of cold and fog make each vintage—and the resultant wine—different."

And so the panels go, explaining each step: "In the California vineyards, a single average vine of the best varieties will bear 12–18 pounds of fruit each year, equivalent to 4–6 bottles of wine."

Some of the information may be a bit too detailed to be of interest to all visitors: "Very heavy pressing yields coarse wine." But the glory of self-guided tours is that you may simply bypass panels that bore you, whereas it's often hard to tune out a boring tour guide.

Following the trail of panels leads you through the winery complex. Each one is appropriately located in an area where you can view the machines and equipment that are described in the panel. A catwalk takes you to the back of the winery, where you can view the stainless-steel tanks and read about the fermentation process.

Two panels on the making of red and white wine give some of the most complete information to be found on a tour about how these two different wines are treated. For example, the panels describe the number of days the different grapes are fermented and the sugar and temperature levels they need during fermentation.

For a lighthearted touch, most panels include quotes from

famous people, or just quotable ones. For example, there is this one from the Greek playwright Aristophanes: "Quickly, bring me a beaker of wine so that I may wet my brain and say something clever."

From the fermentation tanks to the bottling room, you may get the opportunity to see the winemakers in action. The tour concludes on a deck with a view of, and a panel on, the Napa Valley. Then it's time to do some sampling.

The tasting room, opened in 1973, was built in true California style. Part of the room has a glass ceiling, potted trees, and lawn furniture. Another area has paneled walls and wooden tables and chairs for the sun-shy. You help yourself to your first glass; later the other two wines offered for tasting will be served to you.

Some may grumble that for a $5 ride more than three wines should be offered. Visitors to the winery do receive a $2 discount on any wine purchased, however. The wine is not the least expensive in Napa Valley. (And the less-expensive wines are the ones available for tasting.) Prices include $6.25 for a 1987 Cabernet Blanc, $14 for the Winery Lake Pinot Noir, $15 for the 1989 Chardonnay, $16 for the 1987 Diamond Mountain Ranch Cabernet, and $40 for the Sterling Reserve. Sterling concentrates on five varietal wines: Cabernet Sauvignon, Chardonnay, Merlot, Pinot Noir, and Sauvignon Blanc. The varieties are grown in fourteen vineyards throughout the valley, totaling some 1,200 vineyard acres.

Both the tour and the tasting give you the freedom to spend your time as you like, leisurely or quickly. So unless you're full of questions that only an experienced tour guide could answer, the Sterling tram, tour, and tasting is a definite not-miss in the Napa Valley.

DIRECTIONS: From Napa, take Highway 29 north past St. Helena on the way to Calistoga. Watch for both the winery's sign and the road sign for Dunaweal Lane. Turn east on Dunaweal Lane for .5 mile; winery is on the right.

HOURS: Self-guided tours and tastings available from 10:30 A.M. until 4:30 P.M. daily.

EXTRAS: Wine and wine accessories sold at gift shop.

COLORADO

Colorado Cellars, Ltd.
Plum Creek Cellars

In the early 1970s the Four Corners Commission, a government experiment, was launched to determine the feasibility of making wine in the area where Utah, Arizona, New Mexico, and Colorado meet. The results showed that grapes could in fact be grown in this area. Working with the commission, Dr. Gerald Ivancia encouraged the production of wine grapes. When his winery, Ivancia Wine Cellars, shut down in 1975, however, there was no winery left in Colorado to sell the grapes to. Jim and Ann Seewald, home winemakers for years, decided this was the time to go commercial.

According to Jim Seewald of Colorado Cellars, "The Four Corners Commission was the reason for the original plantings. They had proven the viability of the vinifera." In 1978 Jim and his wife established their winery. Before long two additional wineries opened, and Colorado winemaking was on the move again.

Momentum was building until the winter of 1978–1979, when Colorado got hit with one of the coldest winters of the century. Vines froze and died back to their roots. In the following years vines recovered and small new vineyards were planted.

Although grape growers still fight the sometimes deadly winters, Jim Seewald says the acres of vines in the state have increased to almost 250. But almost half of the acreage belongs to hobbyists or doesn't produce high yields.

Three microclimates on the western side of the state are producing vinifera grapes: Grand Valley–Mesa County, Paonia-Hotchkiss, and Canon City–Penrose. These areas have the hot days and cool nights needed to grow vinifera. Much of the acreage in these areas, however, is already planted in peaches and cherries.

It appears that the state's wine industry will continue to grow. In November 1991 the Grand Valley area was federally approved as a viticultural district. Although it may be a long time before people say "Colorado, you mean the wine country?" the dedicated wine growers in Colorado are working on making it happen.

Colorado Cellars, Ltd.

3553 E Road
Palisade, CO 81526; (303) 464–7921

"It grew like topsy-turvy," says Jim Seewald of his home winemaking. It grew so large he went commercial and then became the winemaker for the oldest operating winery in Colorado, Colorado Cellars.

"We started in Golden in 1978. We produced wine there for three years," says Jim. During the summer of 1981, the Seewalds moved their operation to the western side of the state, to Palisade in the Grand Valley. "[We] needed to be where the grapes were grown. It's tough to do at a distance, particularly over the Continental Divide." Jim says it was a hard decision. The smaller population base in the west meant fewer customers.

Jim will be happy to give you a tour of where the grapes grow. "It's not a preset video tour. Or push a button like in the national parks and they tell you what you're looking at," says Jim. His tour covers more than winemaking. "Up until 1921 it [the Colorado River] was called the Grand River. Every other state called it the Colorado River, but there was another little river called the Colorado," says Jim, so Coloradans called it the Grand. From the tasting room, a pinkish stucco building with brown trim, you're given a wonderful view of the Grand Valley, a valley that Jim says has been good to the vines.

"The fact is that with the altitude, the cool nights, we can do a good job with vinifera. We have some problems—every vineyard in the world has. Leafhoppers are our biggest pest," says Jim. This part of Colorado doesn't get as much snow as the eastern half. "I suspect about 4 to 10 inches the whole winter," says Jim. "Our winter temperatures necessitate we grow only winter-hardy."

One of the biggest problems Jim has encountered is not the weather but the lack of grapes. "We don't want to go to California for grapes," says Jim. But he says that until more vineyards are planted in Colorado, there are just not enough grapes to go around.

The winery has gone through several name and ownership changes. It was Colorado Mountain Vineyards, Vintage Colorado Cellars, and Colorado Cellars. Jim owned the winery until 1985; then economics forced him to search for partners. Disagreements among the partners caused a reorganization in 1988, and the winery has new owners now, Richard and Padte Turley.

Jim continues to make the wine. He makes a Chardonnay, Gewürztraminer, Merlot, Cabernet Sauvignon, White Zinfandel, Late Harvest Riesling, and Grand Gamay. And Colorado Cellars makes two kosher wines: Kerem Ram Dry Cherry and Kerem Ram White Riesling. The wines sell for $5.75 to $20 for some of the older Cabernet Sauvignons.

Jim says Rick Turley has broadened the winery's product line to include something to satisfy everyone. The winery has introduced a carbonated cherry wine called Cheré and two champagnes made in the traditional *méthode champenoise* style. Cheré sells for $13, the Riesling champagne for $17.50, and the Cuvée for $20.

Another wine produced is the White Riesling. "Every vintage of that has won medals," says Jim. "I think it's the climate. You get that crisp, clean acidity. We get a lot of German tourists, and inevitably they buy a bottle to take back to Germany."

Germans won't have to take the wine back now. Rick has not

only expanded the winery's products; he has also expanded the markets. Colorado Cellars wine is now sold in Europe and Japan. Rick said in 1992 the European markets tripled their orders.

But you don't have to go to Europe to sample the noncarbonated cherry wine that Jim is fond of. It's made with local organically grown fruit. "If you want to get really elegant, serve it with French vanilla ice cream and almond cookies. Not a macaroon," says Jim, as if using a macaroon would be short of criminal. "It has a cinnamon nose and an almond finish. Very elegant," says Jim, with a look of pure heaven on his face. You can tell he enjoys his work.

And he will share that enjoyment and enthusiasm on a tour of the winery. Jim will answer any of your questions about winemaking and also discuss each wine you taste. "We let them taste anything we have," says Jim.

Watch what you say about the wine. Remember, you're talking to the winemaker.

DIRECTIONS: From Interstate 70, drive south on Elberta Avenue in Palisade for .8 mile. Where it dead-ends into West Eighth, turn left and drive for .7 mile. Turn right at the EAST ORCHARD MESA sign and drive for 4.7 miles. At E Road turn left; winery is on the right at the top of the hill.

HOURS: Tours and tastings offered May to December, Monday through Saturday from noon until 4 P.M. From January to April, the winery's open Friday and Saturday only, from noon until 4 P.M. Tours other times by appointment. Closed New Year's Day, Thanksgiving, and Christmas.

EXTRAS: Winery sells wine, wine accessories, T-shirts, and gift boxes and baskets. Picnic facilities available.

Plum Creek Cellars

3708 G Road
Palisade, CO 81526; (303) 464–7586

Plum Creek Cellars didn't ease into the wine business. Instead, after being bonded in 1985, it went right out and won a silver medal in the Southwest Wine Competition with its 1986 Chardonnay.

"It's all pioneering-type work," says Erik Bruner, winemaker at Plum Creek. "The challenge is there every year, because the grape comes in different every year. We haven't found the definitive style."

Erik began his pioneering by drinking. "When I got married my wife and I got into drinking wine. Then I grew grapes in the backyard and found I had a knack for it. I went to short courses and got some schoolroom education."

Erik and the other owners began in 1983 by planting vineyards. The winery's vineyards are near the Palisade and Peonia-Hotchkiss areas in western Colorado. Their biggest problems have been hail, frost, and intense cold. "The temperature gets to zero or below, and you start getting winter kill," says Erik.

Another problem has been that "the best areas for grapes are where people want to build houses. There's not that many places in Colorado where you can grow wine grapes, especially vinifera."

The winery was south of Denver in Larkspur until 1990, when the owners moved the winery across the Continental Divide to Palisade to be closer to the grapes. "It didn't make sense to carry the grapes over the mountains," says Susan Knisley, who works at the winery. "We process much fresher grapes." Located atop a mesa, the winery is surrounded by five and a half acres of vineyards.

In 1991 Erik's 1990 reserve Chardonnay was named wine of the year by the *Southwest International Wine & Food Review.* The grapes for the Chardonnay came from the thirty-acre vineyard on Sunshine Mesa. At 5,700 feet above sea level, it is one of the highest Chardonnay vineyards in the world.

The winery specializes in Sauvignon Blanc, Chardonnay, Merlot, Cabernet Sauvignon, Pinot Noir Blanc, and Semillon. You can try any of the wines, which sell for $6.50 to $13.50.

A tour, including the tasting, can last twenty minutes to an hour, depending on your interest. During the fall you may enter the winery and watch the staff work the destemmer-crusher or bottling line. If a group is interested, the guide will walk through the vineyards and explain some of the work needed to keep the vines in shape.

If you can't make the trip to Palisade but find yourself in the Denver area, you may still taste the winery's product at its second location. Plum Creek Cellars at 1588 South Pearl Street (303–399–7586) offers samples of its own wine as well as wine from three other Colorado wineries. The tasting room is open from noon to 6 P.M. on Fridays and Saturdays.

The Denver location's L-shaped room has large storefront windows that give the oak floors and handmade oak bar plenty of light. Colorado-made rugs and artwork decorate the walls that aren't covered with wine racks. As you taste the wine, you'll receive information about how the wine is made and also an update on the wine and grape business in Colorado.

That's how Erik and the people at Plum Creek see their jobs: getting people used to Colorado wine and making a product that they will enjoy. When you visit the winery and taste the wine, you'll find they have the second part licked.

DIRECTIONS: From Denver, drive west on Interstate 70. Eleven miles east of Grand Junction, take exit 42. Drive one mile south and road dead-ends into G Street. Turn west and drive .5 mile to winery.

HOURS: Tours and tastings available May through September, Thursday, Friday, and Saturday from 1 P.M. until 5 P.M.; October through April, Friday and Saturday from 1 P.M. until 5 P.M. Other times by appointment. Closed Christmas and New Year's Day.

EXTRAS: Wine and wine accessories sold at the winery.

CONNECTICUT

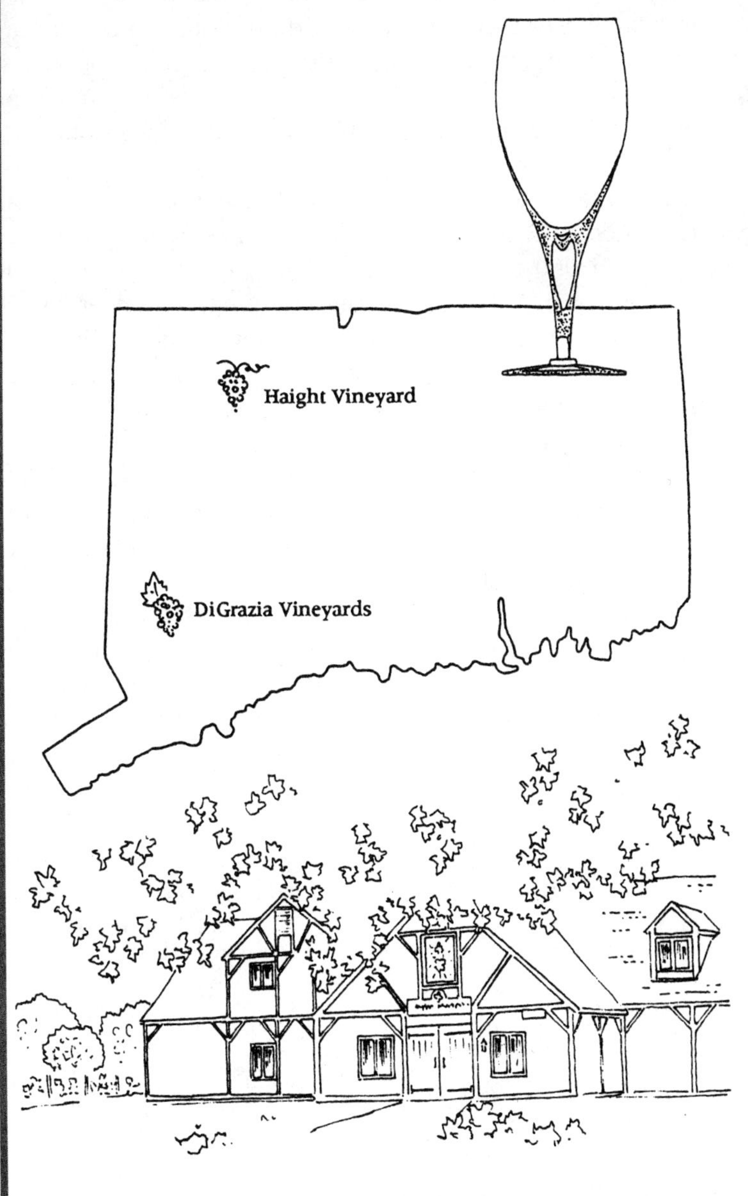

Grapevines and ripe grapes decorate the state seal of Connecticut and symbolize a crop once important to the state. "Before Prohibition, grape growing was the fourth-largest crop," says Barbara DiGrazia, co-owner of DiGrazia Vineyards in Brookfield, Connecticut. But Prohibition-era laws shut down winemaking until the 1970s.

It took the 1978 farm winery bill to make it possible for small wineries to open. Before the bill was passed, wineries had to pay more than $1,000 in licensing fees. The 1978 bill reduced the license to less than $200 for wineries that produced under 75,000 gallons a year. The bill also allowed wineries to offer tastings and sell wine at their tasting rooms. Further legislation in 1987 raised the amount of wine a farm winery could produce from 75,000 gallons to 100,000 gallons.

The 1987 bill also included the formation of the Connecticut Farm Wine Development Council. The council works to promote the state's wineries and to support the wineries with research about grape growing in the state.

Because the laws have only recently changed, there's just a handful of wineries in the state. And winemakers are only just beginning to realize the potential of the area. It's like "flying by the seat of your pants," says Barbara. But it also has its advantages. "The winemakers in our state are very, very close. We call each other up at seven in the morning. We share a harvester, corkers. It makes it really nice."

So far the state's winemakers have had the best luck with white wines, but they hope that with further research and experimentation reds may also find their place in Connecticut bottles.

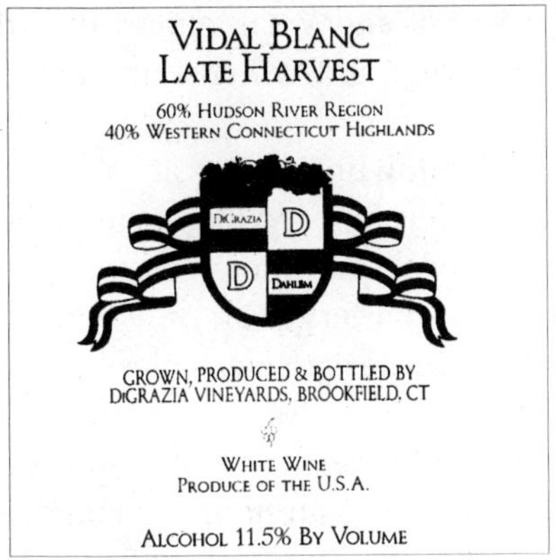

DiGrazia Vineyards

131 Tower Road
Brookfield Center, CT 06805; (203) 775–1616

The DiGrazias do all they can to make sure their winery represents the area. "The names, we think, reflect a New England image," says Barbara DiGrazia, who owns the winery with her husband, Paul. Some of the DiGrazias' twelve wines include Wind Ridge and Meadowbrook, semidry whites; Heritage, a white dessert wine; and Honey-Blush, made with clover honey and grapes. A spicy port called Blacksmith is named after a man who lived in the area all his life and had worked as a blacksmith. "He was here before there were roads," says Barbara.

"We grow Connecticut grapes. We look for varieties that will grow and make good wine," says Barbara. Since winemaking is relatively new to the area, there have been a few trials, with errors. "We tried to produce Chardonnay, but the canes split," says Barbara of a variety that has woody stems that aren't very winter hardy.

It wasn't long before the DiGrazias, licensed in 1984, found grapes that could survive and thrive in the Connecticut climate. "We would rather not fight . . . with nature," says Mark Langford, the winery's business manager. He says using the French-American hybrids, especially Seyval and Vidal, allows the winery to achieve its goal of "grape growing and winemaking aimed at quality and excellence."

The winery's Meadowbrook Classic, known as an ice wine because the grapes are left on the vine to freeze before they're fermented, won a gold medal from the prestigious New England Wine Competition. It's not made every year; only when the weather conditions are right. Other DiGrazia wines, priced from $6.99 to $11.99, have also won awards.

You can stop by and taste the wines and take a tour, too. The tours vary from ten minutes to an hour, depending on your interest. "We have a video, and people can see it [winemaking] actually happening," says Barbara. "No question goes unanswered. On the weekends the winemaker [her husband] is available to talk. He loves to talk," says Barbara with a smile.

During the week Paul attends his other job. "He has a full-time medical practice, so he has two full-time jobs," says Barbara. "My husband went to medical school in Switzerland. He developed a romance for vineyards." It's obvious that the DiGrazias have a romance with wine and Connecticut. When you visit the winery, you won't be able to keep it from rubbing off on you.

DIRECTIONS: From Danbury, drive east on Interstate 84 to exit 9, Route 25. Drive north on Route 25 for 3.7 miles. At Route 133, turn right. After 1 mile turn right on Tower Road; winery is on the left after .1 mile.

HOURS: During January, February, and March, tours and tastings offered Saturday and Sunday from noon to 5 P.M. From April to December, tours and tastings Wednesday through Sunday from noon until 5 P.M. Winery closed New Year's Day, Thanksgiving, and Christmas.

EXTRAS: Winery sells wine, wine accessories, and local crafts. Picnic facilities.

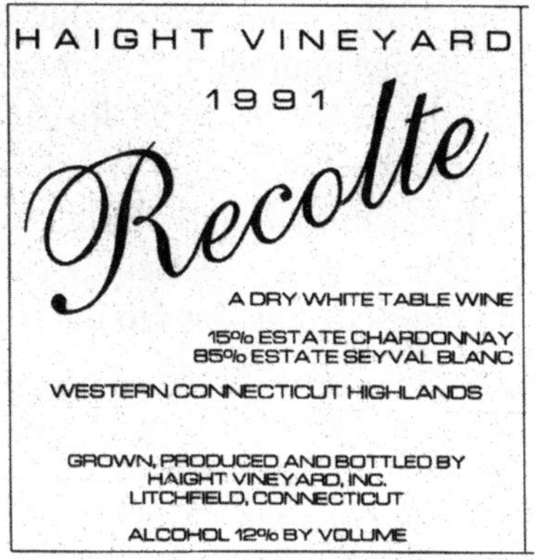

Haight Vineyard

29 Chestnut Hill
Litchfield, CT 06759; (203) 567–4045

Litchfield, home of the Haight Vineyard, is proud of its history. In the early 1700s it was settled by English colonists. Years later the colonists participated in the American Revolution in a most unusual way. They destroyed a statue of King George III and brought it to the farm of local resident Oliver Wolcott. There they melted the statue down to make bullets to use against the British. Wolcott later became one of the signers of the Declaration of Independence. When you visit Litchfield, you can visit his house and the houses of Harriet Beecher Stowe, author of *Uncle Tom's Cabin,* and Oliver Wolcott, Jr., who later became a secretary of the treasury.

A different kind of history was made more than two hundred years later when in 1978 Haight Vineyard became the first commercial Connecticut winery to open since Prohibition. Much like the bang the colonists made with their homemade bullets, Haight made a bang in the wine competitions right from the beginning.

Sherman P. Haight, Jr., the owner, won awards with some of his varietal wines such as Maréchal Foch and Chardonnay. Blends include Chestnut Hill White, a sweet Riesling-style wine; Barely

Blush, a slightly sweet, fruity wine; and Recolte, a dry white table wine. In 1983 Haight claimed another first by making the state's first sparkling wine. The wines sell for $9.98 for the Chardonnay to $11.98 for the sparkling wine.

You may also taste Haight wines in Mystic, Connecticut. Located just off Interstate 95 at exit 90, Haight calls itself "a wine education center" and offers wall displays to inform the visitor about wine through a self-guided tour.

Back at Litchfield you can walk through some of the thirty-three-acre vineyards as part of the tour. As you stroll through the rows and rows of Seyval, DeChaunac, and other varieties, Sherman has signs posted to offer information about the varieties you're viewing and a history of each. Other information indicates when the vines were planted and what awards the wines produced from them have won.

Inside the two-story winery, which is painted white with black wood trim, the second part of the three-part tour begins. Large color photographs line the wall of the first floor. Captions tell how the vines were first planted and how they must be cared for today.

On the second floor a tour guide will take you on a twenty-minute jaunt from the back deck overlooking the vineyards to the cellar. After the tour you'll be offered a tasting on the second floor.

Whether it's a visit to Litchfield where history was made or a visit to the winery where history is in the making, you'll find history doesn't have to be boring.

DIRECTIONS: While driving north on Route 8, take exit 42. Turn left onto Route 118 and drive 4 miles. At Route 254 turn left; winery is .1 mile on the left.

HOURS: Tours and tastings Monday through Saturday from 10:30 A.M. until 5 P.M. and Sunday noon to 5 P.M. Winery closed New Year's Day, Easter, Thanksgiving, and Christmas.

EXTRAS: Winery sells wine and wine accessories. Picnic facilities.

FLORIDA

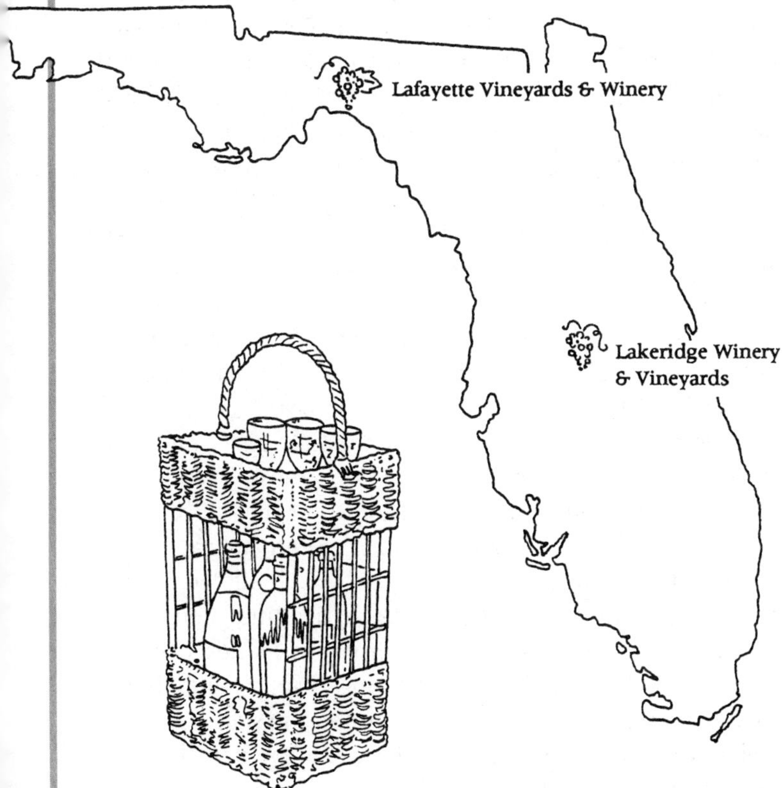

Lafayette Vineyards & Winery

Lakeridge Winery
& Vineyards

When Frenchman Emile Dubois came to North America in the late 1880s, he brought his love for good wine with him. So when he settled in the Tallahassee area, he planted vineyards in several locations and experimented with more than 150 different grape varieties. Emile eventually built a winery, Chateau San Luis, and the wines he produced won silver and gold medals at the Paris Exposition in 1900.

But French immigrants had made wine even before Emile's time. In the 1500s French Huguenots in the Jacksonville area produced wine from wild Muscadine grapes. Also, the Marquis de Lafayette planted grapes and made wine from native Muscadine grapes found on the land around Tallahassee. He had been given that land as a part of a grant for his work in the Revolutionary War.

Central Florida had large vineyards in the late 1800s with northern varieties such as Niagara and Concord. By the early 1900s the amount of acreage in vines had increased 1,000 percent to more than 5,000 acres. Within ten years of reaching that peak, however, most of the vines had been wiped out by Pierce's disease, which attacks nonnative Florida varieties.

The rich tradition reemerges today as land that has gone from grapes to grapefruit once again returns to grapes. According to Florida's Agricultural Research and Education Center, 35 percent of the land that is currently being planted in grapes was formerly in citrus. The state now has 500 to 800 acres planted in grapes.

Florida is one of the few states where the state government has been actively involved in grape research. For almost seventy years the Institute of Food and Agricultural Sciences at the University of Florida has had a grape-breeding program. Grape growers and winemakers believe that their hard work has finally paid off. Dr. John A. Mortensen, a geneticist affiliated with the Agricultural Center, has developed a premium wine grape that thrives in Florida's climate. Named after Emile DuBois, the Blanc Du Bois is still young, though so far it has been well received. Growers around the state, and especially at Florida's Lafayette Vineyards, believe this could be the grape that puts Florida wines on the top shelf.

If the market for wine is any indication, Florida may have a bright wine-growing future. The state has the third-highest volume in wine sales, after California and New York.

Lafayette Vineyards & Winery

Route 7, Box 481
6505 Mahan Drive
Tallahassee, FL 32308; (904) 878–9401 or (800) 768–WINE

It takes a lot of work to run a winery, but who would have thought that you needed someone with physical education experience? That's exactly what Lafayette Vineyards & Winery owes its success to. Jeanne Burgess, a former physical education instructor, is the winemaker at the winery, and she has brought fame to Lafayette with her wines. The winery opened in 1983, and since then it has won hundreds of awards.

Jeanne didn't come right from the locker room to the wine cellar. Her father was a grape grower and winemaker, and she often helped him with his ten-acre vineyard. After studying physical education and having a go at teaching, she returned to her grape roots.

She did graduate work in agriculture at Mississippi State University and worked for another Florida winery before coming to Lafayette.

C. Gary Cox, a general partner, saw Jeanne's potential and brought her in on the beginnings of Lafayette. Cox, a former CPA, organized thirty investors who wanted to bring grape growing and winemaking back to the area. The winery and vineyard sit on the eastern boundary of the Lafayette Land Grant of the 1700s, hence the name of the vineyards. The winery building itself is the perfect combination of the old and the new, with old-world French Provincial architecture and a Florida touch of stucco walls embedded with crushed coquina shells.

The vineyard has Muscadines, such as Magnolia and Welder. Other acres hold Stover, Suwannee, and Blanc Du Bois grape varieties.

The wines include Stover Special Reserve #1, a very dry white table wine; Plantation White, a semidry blended white table wine; Muscadine, in both a semisweet white and red; Blanc Du Bois, a dry white; and Sunblush, a blended wine that is "as rich in intensity and color as a Florida sunset." Lafayette also makes sparkling wine, such as the Blanc De Fleur, in the traditional *méthode champenoise.* Wines sell for $6.95 to $11.95. You may taste the wines in the spacious tasting room with its circular bar.

The winery offers a forty-five-minute tour that includes a slide show and a walk on a catwalk through the winery, which gives you a bird's-eye view of the gleaming stainless-steel fermentation tanks. The catwalk leads to the back of the winery, where a deck provides an excellent view of the vine-covered rolling hills. A window in the tasting room also gives you the chance to watch as Jeanne performs her magic testing the latest batch of wine. It's magic as good as that performed at any winery.

DIRECTIONS: From Tallahassee, drive east on Interstate 10 to exit 31A. Drive west on Highway 90 for approximately 1 mile; winery is located on the south side of the road.

HOURS: Tours and tastings available Monday through Saturday from 10 A.M. until 6 P.M. and Sundays from noon until 6 P.M. Winery closed New Year's Day, Easter, Thanksgiving, and Christmas.

EXTRAS: Winery sells wine and wine accessories.

Lakeridge Winery & Vineyards

19239 U.S. 27N
Clermont, FL 34711; (904) 394–8627

The owners of Lafayette Vineyards built its sister winery, Lakeridge, in 1989. "The area was traditionally citrus, but they had freezes over and over," says Carole Cox, who with her husband and other partners began the wineries. "They're encouraging grape growing in the area. And they're growing pretty well in Florida." Grapes aren't new to Clermont. During the early 1900s almost 2,000 acres were covered with vines.

And of course it doesn't hurt the winery any being near a prime tourist attraction like Disney World. Because of the crowds that come to this part of Florida, the large Spanish-style winery is built for buses and tour groups. But there are places to drink your wine quietly. For instance, you can have a picnic in front of the winery under one of the palms.

Lakeridge and Lafayette wineries share the same winemaker, Jeanne Burgess. But Carole says you can taste a difference in the wine because of the different locations of the vineyards. You can test your taste buds on a dry white Suwanee, light dry Cuvée Blanc, a

Flamingo blush, Chablis, or the Southern White or Red. Lakeridge also makes a sparkling wine called Crescendo. Prices range from $6.95 to $11.95.

If you're feeling touristy, try the citrus wines, which have subtle flavorings of pineapple, orange, and tangerine. Produced with the Sunshine State Winery label, this wine was initially sold only at tourist stops, but it became so popular that Lakeridge decided to sell it, too.

The guided tour at Lakeridge begins with a slide show describing the history of grape growing and winemaking in the state. The group then sees how Lakeridge makes its champagne. Moving outside of the winery, a deck provides a view of the vine-covered rolling hills while the guide explains how the vines are tended. Lakeridge already has thirty-four acres of hybrid bunch grapes, such as Stover, Suwannee, and Blanc Du Bois, and plans to add more.

Like its sister winery, Lakeridge has a catwalk so visitors may walk through the building and enjoy a bird's-eye view of bottling or labeling, whatever needs to be done that day. The guide then leads the group back to the tasting bar, where you may sample three to five wines while receiving information about each.

DIRECTIONS: From Orlando drive thirty minutes west on the Florida Turnpike to exit 285, Highway 27. Drive south on Highway 27 for 2 miles to the winery.

HOURS: Tours and tastings Monday through Saturday from 10 A.M. until 6 P.M. and Sunday from noon to 6 P.M. Closed New Year's Day, Thanksgiving, and Christmas.

EXTRAS: Winery sells wine, wine accessories, books, and gift baskets.

GEORGIA

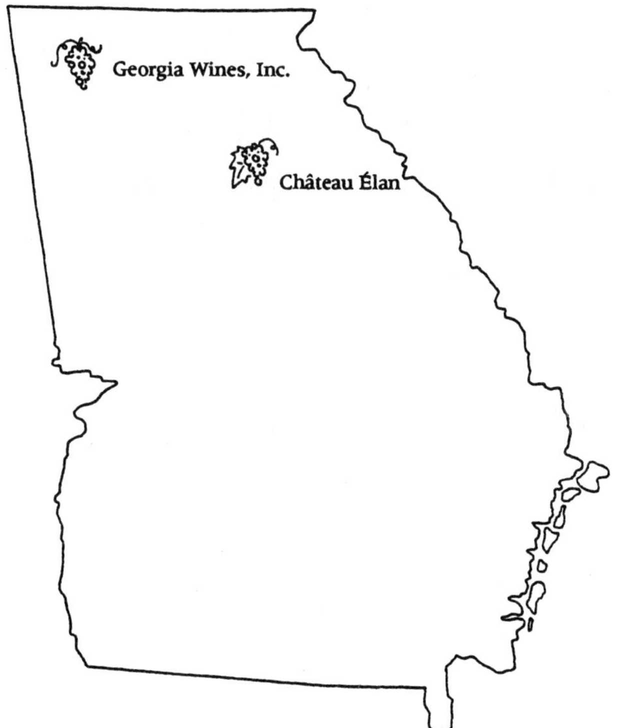

When early explorers came to the area that is now called Georgia, they found an abundance of native grapes, such as the Muscadine. Later, in the 1700s, European immigrants arrived bringing vines with them from their homelands to plant in coastal areas, perhaps believing their vines would adapt. But the European vines were not used to the North American diseases and the hot humid weather. Consequently, most of these early efforts failed.

Winemakers cultivating the Muscadine, however, found they could make a fruity wine with the native grape. Using this grape, Georgia became the sixth-largest wine-producing state by the 1880s. Georgia's wine glory was cut short when the residents voted the state dry long before other states accepted the idea of Prohibition. Commercial winemaking came to an abrupt halt in 1907.

Like many states in the 1970s and 1980s, Georgia has witnessed a rebirth of commercial grape growing and winemaking. This has been prompted, in part, by Georgia's Farm Winery Act of 1983. The act allows wine tastings and the sale of wine at wineries as long as 40 percent of the fruit used in the wine production is grown in Georgia.

The two Georgia wineries included here are as different as red and white wine. Savor the uniqueness of each.

Château Élan

7000 Old Winder Highway
Braselton, GA 30517; (800) 233–WINE

It's not just a winery it's an adventure. Most places would be content with just making a good wine, but not Château Élan. This 2,400-acre, $40-million investment contains vineyard, winery, restaurants, golf course, driving range, petite château golf villas, residential estates adjacent to the seventeenth and eighteenth fairways, retail shops, picnic grounds, nature trails, and meeting rooms. All these features are to be found in and around a sixteenth-century-style French château, which has a pavilion for concerts, a health spa, and more.

The Élan "experience" begins as you drive up the long grapevine-lined road to the top of a hill. The château is surrounded by manicured grounds. As you walk inside the building you find yourself in an open-air French street market, complete with street lights, cafes, and street carts. The large main floor is divided into a gourmet restaurant on your left and a wine market on your right.

Wind your way through the wine display and you're at the corner where the tour begins; you can't miss the 10-by-50-foot mural that illustrates the history of winemaking. With more than a hundred figures, the mural includes scenes of Egyptians making wine and women wrapping rope around bottles, such as those used for Chianti.

In thirty minutes you receive a brief description of how wine is made, a walk through the state-of-the-art winemaking facilities, and a wine tasting.

More than 200 acres of vines provide the grapes for the winery. Château Élan's Chardonnay, Sauvignon Blanc, Johannisberg Riesling, Cabernet Sauvignon, Merlot, and Cabernet Franc have won more than 150 awards.

It also produces wine from Muscadines, the native grapes. Summer Wine has a peach flavor, and the Autumn Blush a raspberry one. Prices for the wine grown and made in Georgia range from $5.50 to $17.25.

You may want to plan your visit around one of Château Élan's special events. There are outdoor concerts throughout the summer, a French Festival in July, and a Vineyard Run in the fall. Check with the winery for the special dates.

At Château Élan you might want to bring your golf clubs or your swimsuit along with your wineglass.

DIRECTIONS: From Atlanta, drive notheast on Interstate 85 about 30 miles. Take exit 48, turn north, or left; the château is located on the hill.

HOURS: Free tours and tastings daily from 10 A.M. until 10 P.M.

EXTRAS: Winery sells wine, wine accessories, T-shirts, and books. Picnic facilities.

Georgia Wines, Inc.

Route 3, Box 900
Chickamauga, GA 30707; (404) 937–2177

A down-home, family-run operation, Georgia Wines will give you all the personal attention you could want. Dr. Maurice Rawlings makes the wine at the vineyard, and his wife, Martha, keeps busy selling it at the winery.

Each fall Georgia Wines has a Grape Stomp. This stomp features music, bonfires, hayrides, wine tastings, winery tours, hors d'oeuvres, and, well, grape stomping. For purple feet, show up at the winery the last Saturday in September.

The Rawlings planted eleven acres of vines in 1983. Although the number of acres remains the same, production has almost doubled. From their vines the Rawlings produce Muscadine Rosé, Muscadine Champagne, Cabernet Sauvignon, Zinfandel, Chattanooga Blush, and twelve other types of wine. They also produce an award-winning fruit wine, Georgia Peach. Wines cost from $4 to $15, with most wines priced at $6.

The wines have won numerous awards, and Rawlings attributes that to his special "freeze fermentation" process, which he invented while working in his lab. He has found a way to keep the yeast nutrient alive and active in cold temperatures, so he ferments the wine at thirty-two degrees Fahrenheit. Rawlings believes that this keeps the flavor of the wine from bubbling away as he says it would during the usual heat of fermentation.

If you are really interested in the winemaking process, bring all your curiosity and questions; they will be well taken care of by people who are in the know. The winery itself, however, is 22 miles from the highway; it's only open for large groups and by appointment. Individuals may sample the wines at the tasting room 300 yards from the interstate.

The tasting room offers wine and lunch. Soups, salads, and sandwiches may be ordered any time during the day. You can also try homemade desserts such as cheesecakes, different types of pound cakes, and strawberry nut cake.

Better still, buy several bottles of wine to drink while picnicking. Then visit the area's attractions, such as Lookout Mountain or the Chickamauga National Military Park.

Lookout Mountain has rock formations and a sweeping view of the Appalachians. It's said the mountain received its name because it was used at one time as a lookout for Indians. The mountain's height ranges from 1,700 to 2,390 feet.

Established in 1890, the 5,400-acre Chickamauga National Military Park is the oldest and largest national military park in the United States. One of the biggest battles of the Civil War was fought on the Chickamauga battlefield. Markers, monuments, and artillery pieces may be found throughout the park. Also, the visitor center at the north entrance on Highway 27 houses the Fuller Collection of American Military Shoulder Arms, with more than 350 weapons.

DIRECTIONS: From Chattanooga, Tennessee, drive south on Interstate 75 and take exit 141; tasting room is on GA2, 300 yards from the interstate.

HOURS: Free tastings Monday through Saturday from 11 A.M. until 6 P.M. Closed major holidays.

EXTRAS: The gift shop at the Tasting Center sells wine, locally made handicrafts, wine accessories, T-shirts, and Muscadine jelly.

HAWAII

Records on winemaking in the Aloha state date back to the early 1800s, when Spaniard Don Francisco de Paula Marin planted a vineyard and made wine on Oahu on land given to him by King Kamehameha I. Don Francisco's wine was served in the royal household and to the king's guests.

Later, early in the 1900s, Portuguese settlers produced wine for their own consumption in Haiku on the island of Maui. They brought the rootstock from Portugal. "There were a number of wineries in the early 1900s, primarily operated by Portuguese, and they used the Isabella grape. The last of those went out of business in the fifties," says Sumner Erdman, the general manager at Tedeschi Vineyards.

He also says that wineries were scattered on almost all the islands throughout the state. "There's no specific growing region. Not like California, where there's a particular area like Napa Valley." Two commercial wineries operated on the islands. One was closed by Prohibition and the other by the Great Depression in the thirties.

By World War II, cattle and pineapples ruled the islands, leaving little space for commercial grape growing. Not until 1974 did commercial wine growing return to Hawaii. C. Pardee Erdman, Sumner's father and owner of a 20,000-acre cattle ranch on Maui, was looking for ways to diversify his enterprise and was considering planting a vineyard. Across the Pacific, Emil Tedeschi, a third-generation Italian living in Napa Valley, had a lifelong dream of running a winery. Mutual friends introduced the two gentlemen, and a vineyard was born.

At an altitude of 1,800 feet above sea level, twenty-two acres on the slope of Haleakala, a dormant volcano in eastern Maui, were selected for the vineyard site. The two men began experimenting with 140 varieties of grapes. The partners chose Carnelian, a vinifera hybrid from Cabernet Sauvignon, Grenache, and Carignane. They began with the idea of producing a red table wine from the Carnelian. But through consultations with the University of California at Davis, the Kula Agriculture Experimental Station, and the world-renowned wine expert Dimitri Tchelistcheff, they decided the grape could make a world-class champagne.

But it would be years before the young Carnelian vines would produce good grapes. In the interim the men decided to produce a pineapple wine to get the winery started. In 1977 Maui Blanc made its debut. The wine sold out in a matter of months—not bad for an interim project. The next year production doubled, and the wine sold out again. Needless to say, this dry fruit wine has remained on the Tedeschi Vineyards' wine list.

The champagne, Blanc de Noirs, was released in December of 1983—the first commercial production of champagne from Hawaiian grapes. Once again Emil and Pardee met with success. The wine was chosen to be served at President Reagan's inaugural ceremony in January 1985, and it received a bronze medal from the Intervin International Wine Competition and a silver from the Atlanta International Wine Competition. This wine is distributed under the Erdman-Tedeschi label, while the other wines bear the Tedeschi Vineyards label.

The Maui Nouveau made its debut in 1985, a Beaujolais Nouveau–style wine that is made to be consumed within the same season it's released. It sold out within months. In the same year, Tedeschi Vineyards added another wine to its list—Maui Blush. The winery hoped to carve out a spot in the growing Zinfandel blush market; in 1989, on Valentines Day, it released Rose Ranch Cuvée, a brut rosé champagne.

The wine industry's future, in the case of Tedeschi Vineyards, appears as bright as the Hawaiian sun.

Tedeschi Vineyards

P.O. Box 953
Ulupalakua, Maui, HI 96790; (808) 878–6058 or (808) 878–1266

Could a winery operate in Hawaii without making a pineapple wine? "That's our main item," says Sumner Erdman about the dry fruit wine. Sumner works as general manager at Tedeschi Vineyards.

"Our main market is tourists. We export one-half our product to Japan, and that's from Japanese tourists who came here." If asked about the advantages of running a winery in Hawaii, Sumner will tell you, "It's the uniqueness. We're not a normal wine-growing region."

The sun doesn't always shine on the winery business on Maui, however, mainly because the sun always shines on Maui. "The big disadvantage is that we don't have a winter and a frost," says Sumner. This doesn't allow the vines to have a dormant period.

But the winery has managed to turn around the disadvantage. "The day we prune is the day spring occurs on the island for that vine." When the vine is pruned it begins its growth and grape production cycle.

Storms are another problem. One nearly wiped out Tedeschi's entire first crop in 1980. The area doesn't get much rain, but when it rains, it pours. "It just comes down in buckets," says Sumner. The vineyard area receives approximately 25 inches of rain a year.

The Maui Blanc, a dry pineapple wine, sells for $6.80. The juice is fermented much as grape juice is and the wine is released approximately three months after bottling. Pale gold with a soft flavor of pineapple, it has received awards at the 1988 Intervin International Wine Competition in Ontario, Canada, the 1988 Atlanta International Wine Festival, and other competitions. The winery sells a quarter of a million bottles of Maui Blanc annually.

The Maui Nouveau, the winery's answer to a Beaujolais Nouveau–style wine, sells for $14.90. The Maui Blush has a light pink color much like a white Zinfandel and sells for $8.85. The champagnes, Maui Brut Blanc de Noirs and the Rose Ranch Cuvée, made in the traditional *méthode champenoise,* sell for $20.15 to $23.

You may sample Hawaii's finest in the tasting room, which was once a jailhouse. On the half-hour tour, you'll learn all about this coral-block building, situated 2,000 feet above sea level, which was constructed in the 1800s by a former sea captain, Captain James Makee. He named the 2,000-acre area Rose Ranch, after the name of his wife's favorite flower, Maui's lokelani rose. While Makee owned the property, he built a sugar plantation and made a fortune selling the sugar during the Civil War. One of his frequent visitors was King David Kalakaua, so Makee built a cottage just for him.

Also on the tour you'll hear about the making of the Blanc de Noirs. "The whole champagne process is explained while you walk among the fermentation tanks and riddling racks," says Sumner.

Apparently owning a winery in Hawaii has its advantages. "It's a beautiful location," Sumner reflects. "Not just the winery, but Ulupalakua itself."

DIRECTIONS: From Kahului Airport, take Highway 37 southeast for 33 miles to the winery.

HOURS: The tasting room is open daily from 9 A.M. to 5 P.M., and guided tours operate from 9:30 A.M. to 2:30 P.M. Winery closed New Year's Day, Thanksgiving, and Christmas.

EXTRAS: Winery sells wine and wine accessories. Picnic facilities.

IDAHO

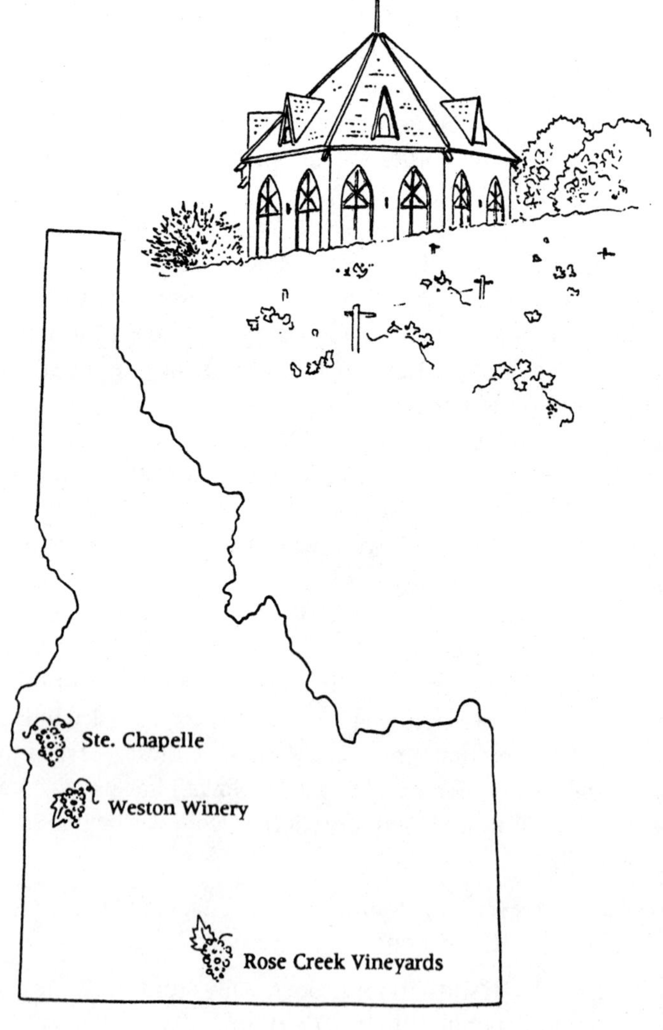

Wine grapes haven't been around in Idaho as long as potatoes, but if the winemakers have any say, they will soon be as famous. The climate is right—similar to that of the grape-growing regions of Oregon and Washington—and the latitude is the same as that of the French wine regions. The vineyards in the Snake River area enjoy long sunny days and rich volcanic soils. Water for irrigation comes from the Snake River or the Lost River Aquifer.

Wineries came to Idaho in the late 1970s. Climate aside, conditions were perfect, with the increasing interest in U.S. wines, the rise of wine consumption in the United States, and the change in state laws that made it possible for wineries to serve wine out of their tasting rooms. Idaho's first winery, Ste. Chapelle, is probably also its foremost, if for no other reason than its size and ground-breaking work in the state. Since its first crush in 1976, Ste. Chapelle has brought the state recognition for its award-winning Chardonnay and Pinot Noir. The eighties brought more wineries to Idaho, and the state's success has continued. As more vineyards mature, along with the state's winemakers' experience, more recognition will come to Idaho for its wine.

Rose Creek Vineyards

P.O. Box 356
111 West Hagerman Avenue
Hagerman, ID 83332; (208) 837–4413

James G. Martin, the father of Rose Creek's owners, pours the perfect amount of wine for tasting. It's neither too skimpy to appreciate the wine nor too much to finish completely before the next taste. James helps out at the winery while his daughter, Stephanie, makes the wine, his daughter-in-law, Susan, manages the business, and his son, Jamie, handles the promotion.

The winery is in the basement of a building that is more than a hundred years old. Built in 1887, it was a stagecoach stop and a store and now has a bank operating in the upper floors. The basement has 3-foot-thick stone walls of native lava. "We took a blower and blew them out, regrouted them, and shored up the roof," says James.

A spring runs under the building, keeping it cool. "You have no fluctuation in the temperature, which would ruin the wine," says James. Black walnut counters and wrought-iron doors decorate the tasting room.

Pictures of the winemaking process are part of the tour, and James will explain what you are looking at. "It's an Italian press, free flowing. A lot of people will crush [the grapes] again. We don't." He will also talk about the soil and the climate. "We have cold nights and hot days, and that puts the natural sugar in." If you visit during harvest in October, the tour will include a trip to the crush pad and an offer to taste fresh juice.

Vines were first planted in 1979. The vineyards have increased from three to thirty acres and production from 1,900 gallons to 10,000 in 1990. The "official" tour is about fifteen minutes, but leave plenty of time for questions and general wine talk.

Then it's time for the tasting, and James will give you more information about the wine and the state: "Idaho is becoming known as a Chardonnay state. All French oak. It gives it the flavor people want who like dry wine. We're fermenting that right in the oak."

Stephanie says there's still a lot of work to be done. "We're finding what grows the best here. We're not going to grow marginal grapes." Most of their red grapes froze in 1989 and 1991, so the Martins are looking to hardier varieties. They have planted Merlot along the Snake River and believe the grapes will do well in the region.

After they get the grapes to grow well, they have to protect them. "Birds are our problem," says James. "I think they come from all over the country just to eat our grapes."

What the birds don't eat becomes Riesling and Chardonnay. The winery also makes wine from other states' grapes: Oregon-grown Pinot Noir, and Washington-grown Cabernet Sauvignon. Prices range from $4.75 to $12.99. Rose Creek's Chardonnay, Riesling, and Pinot Noir have all been award winners.

"People come here for every other reason than wine—the trout, the smoked fish. It's ideal to be on the Thousand Springs Scenic Route," says Stephanie. Rose Creek offers smoked fish, alone or in gift boxes. Smoked red trout, coho salmon, pheasant breast, rabbit, and teriyaki smoked prawns may be combined with the wine for a gift pack for a special friend or as a special present to yourself.

Even though Stephanie says people aren't rushing to the area solely because of the winery, she and her family still enjoy it. "It's fun to be in on it because it's kind of new. One day Idaho will be known for Chardonnay, not potatoes."

DIRECTIONS: From Twin Falls, drive west on Interstate 80 until you reach the Bliss exit. Travel 8 miles south to Hagerman. On the south end of town, the winery is on the right, in the basement of the Idaho State Bank. Also from Twin Falls, you may drive along Highway 30 to Hagerman, along the Thousand Springs Scenic Route.

HOURS: Tastings and tours available daily from 11:30 A.M. until 5:30 P.M. Closed New Year's Day, Easter, Thanksgiving, and Christmas.

EXTRAS: Winery sells wine, wine accessories, local food products, and T-shirts.

Ste. Chapelle

14068 Sunny Slope Road
Caldwell, ID 83605; (208) 459–7222

The winery is named for the thirteenth-century church, La Sainte-Chapelle, or the Saints Chapel, on the Île de la Cité in Paris, built by King Louis IX. The tasting room keeps the feeling of La Sainte-Chapelle with two-story windows, vaulted ceilings with wood beams, and stained-glass windows, only here the windows depict grapevines and grapes instead of religious figures and events.

With sales in more than twenty-five states, Canada, and Taiwan, and ranking as the sixty-sixth largest winery in the United States, the winery's size is impressive. What the owner, Symms Fruit Ranch, did for growing and shipping peaches and apples it is also doing for grapes.

Resting on Winery Hill in an area known as Sunny Slope, the winery, through irrigation from the Snake River, has turned brown hills and valleys into lush greenery. You can enjoy the view from the picnic grounds and gazebos that overlook the Snake River Valley.

The twenty-minute tour begins under a picture of the winery's namesake, the Saints Chapel in Paris, and includes a walk through the bottling room, crush area, laboratory, cellar, and barrel room. You'll see photographs of how oak barrels are shaved and toasted to give wines their oaky flavors, something not usually found on a tour.

Then it's back to the tasting room, unless you decided to skip the tour and immediately taste. You'll receive four tastes, perhaps a Chardonnay, a Riesling, one of three blushes, and a dessert wine or sparkling wine. The staff will guide you through the tasting to find the wines you're most likely to favor.

Ste. Chapelle makes a Special Harvest Riesling, Johannisberg Riesling, Dry Riesling, Chardonnay Reserve, Cabernet Sauvignon, Merlot, Chenin Blanc, Soft Chenin Blanc, Pinot Noir Blanc, Pinot Noir, and four sparkling wines. Most wines sell for $6 or $7. Older vintages, such as the 1986 Chardonnay or 1981 Cabernet Sauvignon, cost more, up to $24.99.

The Winery has a "Jazz at the Winery" series every summer in July. Gene Harris, Idaho's Grammy-award nominee jazz pianist, has been the main attraction of the Sunday series. You may enjoy the music, purchase wine by the glass or bottle, sip soft drinks, and munch on freshly baked bread and a variety of cheeses. The latest list of concerts and prices may be obtained by writing or calling the winery.

For those who choose to stay in Boise, Ste. Chapelle has a tasting room there, too, Ste. Chapelle on the Grove. Located at 801 Main Street (208–344–9074), it offers tastings and wine sales Saturdays and Monday through Thursday from 10 A.M. until 6 P.M. and Friday from 10 A.M. until 9 P.M.

Ste. Chapelle has some of the most informative labels. After a brief history and some background on the winery, the label provides such information about the wine as release date, alcohol and acid content, taste, recommended serving temperature, and foods to accompany the wine. The Chardonnay label, for example, reads in part: "Complementary foods—steamed clams, grilled Chinook salmon, veal in a mustard-cream sauce, beef filet roasted in herb butter, cream of wild mushroom soup." It's enough to make you grab a bottle from the winery, run home, and start cooking.

DIRECTIONS: From Boise, take Interstate 84 west to exit 35. Follow Highway 55 south, approximately 16 miles. Turn left on Lowell Road; winery is .8 mile on left.

HOURS: Summer tastings offered June through September, Monday through Saturday 10 A.M. until 6 P.M. and Sunday from noon until 5 P.M. Winter tasting hours are October through May, Monday through Saturday 10 A.M. until 5 P.M. and Sunday from noon until 5 P.M. Tours available year-round on the hour, beginning one hour after opening and ending one hour before closing. Limited tastings and tours during the winery's jazz concerts in the summer.

EXTRAS: Winery sells wine, wine accessories, and T-shirts. Picnic facilities.

Weston Winery

16316 Orchard Avenue
Caldwell, ID 83605; (208) 454–1682

"At twenty-eight-hundred feet we have the distinction of being at the highest altitude in the Northwest," says Cheyne Weston, owner of the Weston Winery. According to Cheyne, his high-altitude, ten-acre vineyard has many advantages. "It usually brings a more intense fruit flavor to the wine. Being at a high altitude, the Riesling does real well for us." Both Weston and Ste. Chapelle have won international and national awards for their Rieslings.

The soil is sandy and volcanic, which provides good drainage for the vines' roots. The area has warm sunny days with cool nights, and a very cold winter, which some varieties can't survive. Most vineyard owners believed Cabernet Sauvignon couldn't grow in Idaho, but Cheyne believes he might be able to do it. "It was thought it couldn't grow, but that statement has proved to be incorrect. We're trying to develop the most classic wines."

Cheyne sees it as a learning process. "Idaho is a fairly new wine-producing state." Also new is Idaho's wine commission, which Cheyne was appointed to by the governor.

Weston Winery is both the second-oldest and the second-largest winery in the state. Cheyne worked at Ste. Chapelle for two years before branching off on his own. The wine bug bit him years earlier, however, when as a student he worked in a neighbor's vineyard in his native Oregon.

Since 1981 Cheyne has been growing grapes. His wines are Johannisberg Riesling with "an apple-scented nose," Chardonnay with "their toasty-buttery characteristics," Cabernet Sauvignon "ruby colored and full bodied," Pinot Noir with a "light-perfumed nose," and a "fruity and refreshing" Blush.

"We're the first to make a Pinot," says Cheyne. He also makes sparkling wine in the traditional *méthode champenoise.* "Right now we make 1,050 bottles per year. We're in the experimental stage with it. We also sell out of that in seven weeks." Not bad for an experiment. The wine sells for $5.99 for a Riesling to $15 for the champagne.

All the wine is sold under the River Runner label. Cheyne believes that the river runner reflects the adventuresome, pioneering spirit of grape growing in the Pacific Northwest. Another reason for the label is Cheyne's personal interest in river rafting; in his spare time he works as an oarsman and guide.

A white building with black wrought-iron trim houses the winery and tasting room. The winery produces approximately 5,000 gallons a year. But Cheyne adds, "It's been growing a little bit each year." When you tour the winery you'll see the growth. There are barrels everywhere, filled with aging wine. You decide the length of the tour, depending on your interest.

DIRECTIONS: From Boise, drive west on Interstate 84 to exit 35. Winery is on the right approximately 15 miles south on Highway 55.

HOURS: Tastings and tours daily from 1 P.M. until 5 P.M. Winery closed New Year's Day, Easter, Thanksgiving, and Christmas.

EXTRAS: Tasting room sells wine, wine accessories, and T-shirts. Picnic facilities.

ILLINOIS

Winemakers in Illinois have been faced with two major obstacles. First, wineries could not offer wine samples or sell wine until 1981. Would-be winemakers believe that recent legislation might return the state's wine industry to its former glory, if the second problem can be solved: growing vines in land contaminated by the chemicals that help corn grow.

More than one hundred years ago people were growing wine grapes along the Mississippi, especially around the northern Illinois town of Nauvoo. Nauvoo was settled by Joseph Smith and a group of Mormons in 1839. Persecution and the murder of Smith pushed the Mormons out of Nauvoo in 1846, when they began their trek to Utah.

After the Mormons left for the Great Salt Lake Valley, the French Icarians came in 1849 and tried communal living. When the commune broke up in 1857, many of these French families remained in Nauvoo. With their German and Swiss neighbors, they planted grapes and brought winemaking to Nauvoo. Soon the hills of Old Nauvoo were dotted with vineyards and honeycombed with wine cellars. Nauvoo became famous for its fine wines.

Emile Baxter, the great-great-grandfather of the present owner of Baxter's Vineyards, came to Nauvoo in 1855 to join the Icarians. When the commune dissolved he decided to stay. He bought eight acres of land and planted grape canes obtained from Cornell University.

His winery flourished, and when his sons were old enough, they joined the business to form Emile Baxter & Sons. When Emile died in 1895, his sons continued the winery as the Baxter Brothers. During Prohibition the Baxters continued making wine for family consumption only and shipped their grapes to northern markets.

With repeal, Cecil, a next-generation Baxter, reopened the family business. In 1936 Cecil applied for a license, and his Gem Vineland Company became the first bonded winery in Illinois since Prohibition. The Baxter family continued operating the winery and formed a family corporation in 1956. The winery closed for some of 1986 and 1987, but the present owners, Brenda and Kelly Logan, are trying to keep the Baxter family tradition alive. Under their ownership, the winery returned to its original name: Baxter's Vineyards.

Baxter's Vineyards

P.O. Box 342
2010 Parley Street
Nauvoo, IL 62354; (217) 453–2528

"Supposedly it only happens in two places in the world, Roquefort, France, and here," says Brenda Logan, who with her husband, Kelly, owns the winery. It's the annual "Wedding of the Wine and Cheese," celebrated for more than fifty years during the annual grape festival. Held the Saturday and Sunday before Labor Day, the whole town of Nauvoo celebrates its history of wine growing and the more recent venture into making blue cheese.

The festival culminates with the "wedding." The story goes that a young French shepherd forgot his lunch in a cave one day. Returning months later he found the cheese had molded, leaving blue stripes through the cheese. He found it tasty, and the word spread about the new Roquefort-type cheese. In the Nauvoo ceremony a young boy, representing the shepherd and the cheese, walks down the aisle followed by cheese makers. A young maiden enters with wine. The cheese and wine are placed on a wine barrel, and a judge reads the ceremony from a scroll. Then the scroll, cheese, and wine are encircled with a wooden hoop, signifying the marriage ring. Thus occurs the union of two of the best things in life.

Approximately 14,000 people attend the wedding weekend. Other events of the festival include pancake breakfasts, grape stomping, pet parades, dances, beer gardens, and variety shows. But you don't have to visit Nauvoo only during the grape festival to have plenty to do. Even without the festival, Brenda says, "It would take a good three days to see everything."

The Nauvoo Chamber of Commerce Uptown Tourist Center provides information to help you enjoy your stay. If you like to do things by yourself or with your family, the city's tourist center offers a cassette-guided tour. If you like to travel with a group, the center offers "Step on Tour Bus Guide Service." Both the Church of Jesus Christ of Latter-Day Saints (Mormons) and the Reorganized Church of Jesus Christ of Latter Day Saints operate visitor centers. They offer a movie on the history of the Mormon movement in Nauvoo.

Restored homes and shops have guided tours free of charge. For those who enjoy the outdoors, the Nauvoo State Park offers camping, fishing, and a museum. And the Great River Road, the 3,000-mile road that follows the Mississippi, runs through Nauvoo and has picnic and scenic turnoffs.

Or you can visit the historic winery for a guided tour or a cassette tour. But Brenda would rather give the personal touch, "We prefer to take them." Whether with a guide or cassette, you'll be able to see and hear about the Baxter history. There's an old grape press that Brenda says came from France before 1900 and a unique apple press. While most presses are round, this apple press, dated 1874, is square. "We have three casks that hold twenty-three-hundred gallons. They were built in the room probably right after Prohibition," says Brenda of the casks that are so big they wouldn't have fit through the barnlike doors.

The array of wine is spread out on a counter where you may help yourself to Concord, Niagara, rosé, Sauterne, and Burgundy. "We're known for our sweet wines," says Brenda. All the wines are priced under $5.

Brenda says the most important thing she would like people to know about Baxter's is that "we're here and open, and going again, and would be happy to see everybody."

DIRECTIONS: From Highway 96 driving north, turn right at the Nauvoo State Park entrance. Drive east .9 mile on Parley Street; winery is on the right.

HOURS: Tours and tastings Monday through Saturday from 9 A.M. until 4:40 P.M., and Sunday from noon to 4:30 P.M.

EXTRAS: Winery sells wine and wine accessories.

Waterloo Winery

725 North Market Street
Waterloo, IL 62298; (618) 939–8339

"We're growing on reclaimed coal land," says Mark Hendershot, owner of Waterloo Winery. "That makes us unique." Land once stripped so huge shovels could rip out the coal underneath has been transformed into Mark's Garden of Eden Vineyards. Mark says the strip-mining company has a good reclamation project.

The company breaks the soil, topsoil, and subsoil into various layers, and after the coal is removed, it puts the soil back into place. Mark says the layers are "all mixed up but the minerals are all in it." Then 24 inches of topsoil are put on. "Once we were given energy [from the land], and now we try to do it again with the wine. It's turning out quite well."

Mark says he gets something else from the coal company besides the land. "They have a good reclamation program, and that gets attention." And the attention the company gets overlaps onto his business, which he says, like most wineries, he needs. "You can make the best wine in the world and people won't come knocking at your door. You have to find a way to tell people about it. We're still babies, learning the market," he says.

"This winery here is built into an early farmstead, and it goes back 175 years," says Mark. "We're trying to maintain the farm as it looked in the early 1800s. If there's any expansion, it would be done underground." Mark says one farmer who had the farm many years ago also made wine there. "Historically, Waterloo and all these little towns had wineries all through the 1800s, and they were lost at the turn of the century."

You can recapture a feel of the history while you sit in the front yard of the winery. Turn your back to the McDonald's across the street and relive the history under the old statuesque trees. Waterloo provides the tables and chairs, and Mark will provide the history.

"It's an education. It's a mix of history and winemaking," says Mark of his tour. "It's more a family thing—we don't get motorcycle gangs." Mark believes he offers one of the most interesting tastings in the country, because of his varieties. In 1981, when Mark planted the vines, he planted thirty-five varieties. "I was like a kid under a Christmas tree because I had the opportunity to plant them all." He couldn't resist, even though the number of different vines makes it difficult. Mark says, "It's a challenge and a half."

From the vines he makes sixteen kinds of wine such as Maréchal Foch, Millot, DeChaunac, Ravat, Chambourcin, Riesling, and Diamond. "Only two wineries in the United States are growing that," he says of the Diamond grape. This wide range of varieties allows you to compare the grapes and, in the case of the Diamond, to try a wine you wouldn't usually be able to taste. "It's a tasting education," says Mark. People see what "can be done with grapes." Mark's wine sells for $4 to $10.

A visit to Waterloo Winery, says Mark, means joining him as he "takes time to smell the roses; except instead of roses, we have wine."

DIRECTIONS: From Interstate 255, drive south on Highway 3 for 9 miles; winery is on the right.

HOURS: Tours and tastings offered on Sunday from noon to 5 P.M. from April 1 to December 31. Other times by appointment.

EXTRAS: Winery sells wine and wine accessories. Picnic facilities.

INDIANA

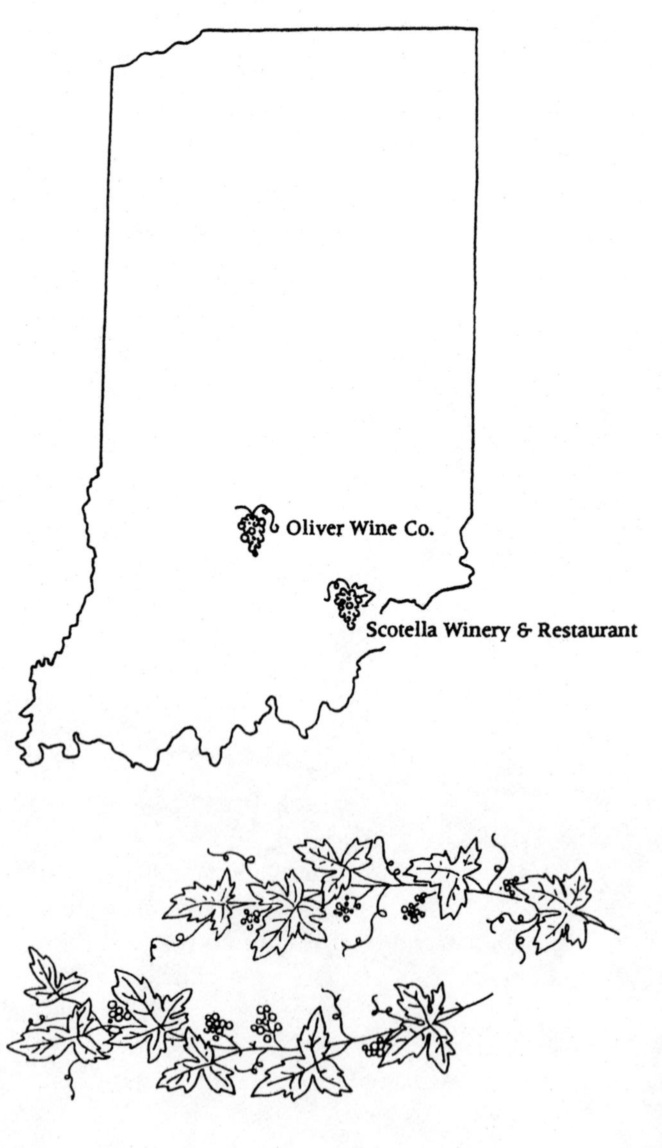

In the early 1800s Jean-Jacques Dufour planted vineyards at Vevay, in Switzerland County. He had worked in vineyards in his native Switzerland and hoped to create a wine industry in Indiana. Noted for having the first successful commercial wineries in Indiana, he also wrote one of the first books about viticulture, *The American Vine-dresser's Guide*. His vineyards, however, did not last long after his death.

More vineyards were planted in other parts of the state with varying degrees of success. According to the Indiana Wine Grape Council, by the mid-1800s Indiana ranked as the tenth-largest wine producer in the United States. The black rot, however, which ruined vineyards in neighboring states soon attacked vines throughout Indiana as well. This fungus, along with the Prohibition movement, shut down the state's wineries in the mid- to late 1800s. It's taken almost 150 years for wine grapes to return.

William Oliver, a law professor at Indiana University in Bloomington, wrote the Indiana farm winery bill and helped it pass in 1971. He then went from home winemaker to commercial winery owner.

The grape growers have had support from state universities. The Purdue University Extension Service organizes a yearly grape and wine symposium and has also established an Indiana Winegrowers Guild.

Since the passage of the farm winery bill, vineyards are being planted all over the state, with approximately a hundred acres under vines. According to the Wine Grape Council, wineries currently produce more than 42,000 gallons of wine in a hundred different styles. And three more wineries are in the planning stages.

One problem that confronts vineyard owners is Indiana's weather. Drastic changes of temperatures have killed off acres of vinifera vines. Some growers are planting the hardier French-American hybrids and native American varieties. The next twenty years will tell whether vineyards in Indiana can continue their comeback.

Oliver Wine Company

8024 North Highway 37
Bloomington, IN 47401; (812) 876–5800

Listening to William M. Oliver, it sounds as if it is still as painful as when it happened. "1983. February. We had a week of warm weather. The grass was starting to turn green. The vines started to bud slightly, and they had a lot of sap in them. We had sixty-degree temperatures go to twenty-three degrees below zero in twenty-four hours. The vines literally exploded. Thirty-five acres wiped out overnight. We don't want to expose ourselves to that again."

So instead of growing grapes, the Olivers, owners of the oldest and largest winery in Indiana, decided to start buying their grapes. But it wasn't always that way. "They [hybrids] grew well," says William. His father, William W. Oliver, planted vines in 1965. "The viability of grapes looked good, and so he contacted the government in the late sixties. Ultimately he wrote the winery act, and that passed in 1971."

William W., a law professor at Indiana University, is well known as the person who cleared the legal path for small wineries in the state. And now his son runs the winery. "He's an attorney and just has too much to do, so that's what I do," says William. "Our goal here is to produce top quality wine from vinifera grapes. It's all hands on and trial and error."

The Oliver wines have done well at the Eastern Wine Competition. The Strawberry won best of class. Another winner both in medals and taste is the Gewürztraminer; the 1989 vintage won a gold medal at the San Francisco Fair. The 1988 Merlot and 1988 Sauvignon Blanc were awarded silver medals at the San Diego National Wine Competition.

Some of the dry Oliver wines are Semillion, Sauvignon Blanc, and Merlot. The semidry wines are Chenin Blanc, Barbera Blush, and Riesling. Semisweet wines include Camelot Mead, Soft White, and Soft Red. Oliver also makes sparkling wine in the traditional *méthode champenoise.* Prices range from $4.76 for Professor Oliver's Peach Wine Coolers, to $6.90 for the 1989 Barbera Blush, to $11.90 for the 1990 Seyval Blanc, and $14.76 for the 1989 Merlot.

The server will tell you in which order the wines should be sampled. After you've found a type you like, you may want to buy a cold bottle along with some cheese and head for the Olivers' pond for serious study. Behind the tasting room and down a little hill, picnic tables are scattered among the trees alongside the pond.

On Saturdays you can take a thirty-minute tour of the complex. Says William of the tour, "It's basically two parts: one on Indiana's history and then the real specifics of how we make wine." William says the Oliver tours are different from those you'll find at other wineries. "We try to really hit on a few high points and how winemaking is really different than it was fifty years ago, with the advent of cold fermentation and sterile bottling. The intense fruitiness and quality wasn't possible before."

William sees a solid future for the Oliver winery. "There's a large section of Indiana that is growing more sophisticated in drinking wine, and we're just trying to follow that group. We're planning a 25 percent growth rate a year." Even though the winery is growing, William doesn't plan on becoming a wine baron. "You open a ball-bearing plant to make money, but you open a winery to be creative."

DIRECTIONS: From Bloomington, drive north on Highway 37 for

6.8 miles; winery is on the right. From Indianapolis, winery is approximately 40 miles south.

HOURS: Tastings offered Monday through Saturday from 10 A.M. until 6 P.M. and Sunday from noon to 6 P.M. Tours available on Saturday or by appointment. Winery closed New Year's Day, Election Day, Thanksgiving, and Christmas.

EXTRAS: Winery sells wine, wine accessories, and picnic foods. Picnic facilities.

Scotella Winery & Restaurant

R.R. 2, Box 2
Madison, IN 47250; (812) 265–3825

Madison is a little gem in southeast Indiana, virtually undiscovered by tourists. Perched on the north side of the Ohio River, this was once the largest city in the state. Almost every building you see in Madison is on the National Register of Historic Places—133 blocks in all. Founded in 1809, Madison features styles of architecture such as Classic Revival, Gothic, Georgian, and Americanized Italian Villa. One of the best times to visit these historic homes is during the "Nights Before Christmas Candlelight Tour."

For the tour, held during the last weekend in November and the first weekend of December, the town decorates the houses in century-old style with lights and greenery. Christmas music and carolers fill the streets as you wander from house to house. "There's a lot to see," says Elsa Conboy, who owns the Scotella Winery with her husband, Scott. "It's a real interesting town."

Some of the most interesting historic buildings are a block or two off Main Street. You can see what medicine was like in the mid-1800s by visiting Dr. William Hutchings' Office and Hospital. His original medical equipment, the house's furnishings, and his hospital are on display from May 1 to November 1, Monday through Saturday from 10 A.M. until 4:30 P.M. and Sunday from 1 P.M. until 4:30 P.M. There's also the oldest Indiana fire station, the Fair Play Fire Co. No. 1, with its horse-drawn fire wagons. Hours vary, so call before visiting.

One old building now houses the Scotella Winery & Restaurant. The building was originally a dairy. "There've been dairies in here since the 1800s," says Elsa. Scott and Elsa bought the vineyard, about eleven acres worth, that was planted in 1978 by the Villa Medco Winery. When that winery went out of business, Elsa says, "My husband and I got interested in the area. We bought it in 1982."

But that didn't mean the Conboys stepped into a ready-made winery. "The vines hadn't been pruned in a few years. We had to walk through with a machete," says Elsa. "We're still getting them back up to snuff. They've come back very well." Four years after the Conboys bought the property, Scotella opened. "We spent a lot of time piecing it together." They also pieced together Scott's and Elsa's names to come up with the winery's name—Scotella.

The Conboys have made the winery a cheery place with white brick walls, vine-covered poles, and a yellow-and-white-striped awning at the entrance. The yellow and white stripes are also found inside; they cover part of the ceiling. Large windows filled with hanging plants have southern exposure to take advantage of the sun all year long.

In 1989 Elsa and Scott opened an Italian restaurant in the winery featuring homemade food from Scott's mother. The entrees include minestrone soup, salad, relish dish, a loaf of bread with cheese, and a sundae with wine or chocolate sauce. Pick from lasagna or spaghetti for $12.95, and seafood neptune or veal piccata for $13.95. Children twelve years and younger may have spaghetti or lasagna with chicken drumstick and a dessert for $5. The restaurant serves food Monday and Thursday from 5 P.M. until 8 P.M., Friday and Saturday from 5 P.M. until 10 P.M., and Sunday from noon to 8 P.M.

Wine by the glass sells for $2, or if you're interested in a bottle, the Baco Noir is $8, the Vidal is $7, and the red or white Duganaire is $6.

If you visit early in the day before things get crazy with the restaurant, Elsa will take you through the winery and explain the winemaking process. "It's basically the one room," says Elsa of the combination fermentation and bottling room. "I will actually take them through step by step."

Spend an hour or two at Scotella, appreciating how a young winery does things and eating homemade Italian cuisine, but save some time for Madison, learning about the past.

DIRECTIONS: From Interstate 65, take exit 29. Drive east on Highway 56 for 24.7 miles. In Madison, drive north on Highway 421 for .6 mile (5 blocks) to Aulenbach Road. Turn right; after three blocks winery is on the right.

HOURS: Tours and tastings available Monday, Wednesday, and Thursday from 11 A.M. until 6 P.M., Friday and Saturday from 11 A.M. until 11 P.M., and Sunday from noon to 8 P.M. Winery closed New Year's Day, Election Day, and Christmas.

EXTRAS: Winery sells wine and wine accessories.

MARYLAND

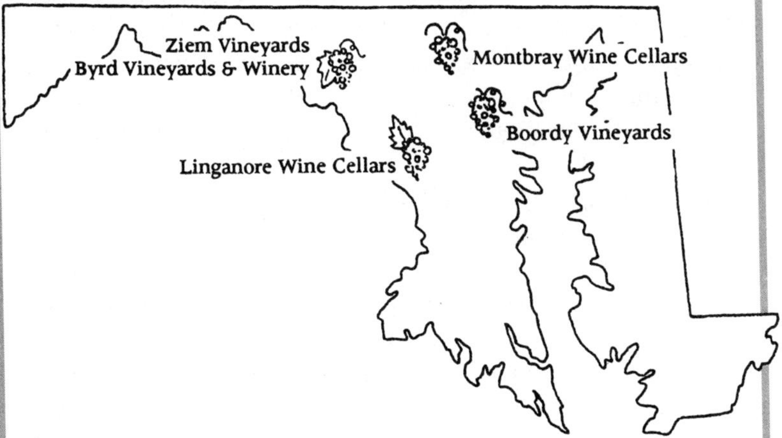

Ziem Vineyards
Byrd Vineyards & Winery
Montbray Wine Cellars
Boordy Vineyards
Linganore Wine Cellars

As early as the 1600s, colonists in Maryland were cultivating grapes and making wine. Lord Baltimore, the English proprietor in the area, tried to establish a vineyard with vinifera in 1662. But his efforts failed because vines from Europe had trouble adapting to the New World. In the late 1770s John Adlum planted grapes in the area now known as Washington, D.C. He named the grape Catawba, and later wrote a book about his experiences. In 1820 the state legislature tried to promote vineyards with the founding of the Society for Promoting the Culture of the Vine.

Not much happened with the Maryland wine industry until many years later. The state will always be mentioned in wine books, however, not for its early efforts or successes but for Philip Wagner, called the "Father of French-American Hybrids." He began making wine as a hobby in the 1930s. Philip had little success with his plantings of California grapes. Looking for an alternative to the native American grapes, Philip opted for hybrids developed in France from American rootstock.

On previous trips to France, he had heard about the hybrids and decided to try them in the state. Philip was pleased with the wine and soon became a great promoter of the hybrids. He spread the word through books and articles, and he spread the hybrids by selling vines from his nursery. In 1945 he went commercial, and soon his wines were in restaurants, promoting themselves. Since then French-American hybrids have made their way up and down the East Coast and west to states like Michigan and Ohio, much to the credit of Maryland's Philip Wagner.

Boordy Vineyards

12820 Long Green Pike
Hydes, MD 21082; (410) 592–5015

"Boordy was started in 1945 by Philip Wagner. The Deford family grew grapes for him," says Julie Deford, wife of Rob Deford, one of the winery's owners. Mr. Wagner founded Boordy after many years of making wine at home growing French-American hybrid plants to sell to vineyards. He has been a big supporter of hybrids and through his articles and books has helped spread them throughout the country.

In 1980 Philip sold the winery's name to the Deford family. The Defords opened Boordy on the family's 250-acre farm. A 19th-century barn houses the winery. The barn is a beautiful building with a stone cellar with hand-hewn beams and wood-planked floors. "He [Rob] and a friend converted the barn to a winery," says Julie, "We've been updating and expanding ever since then."

Part of the updating was Rob's formal wine education. "Rob grew up tending the vineyard," says Julie. But he wanted more training before taking the reins of Boordy. "He rushed through UC-

Davis," says Julie. The University of California at Davis is one of the world's foremost viticultural schools. Although Wagner's Boordy was a long-time champion of hybrid grapes, the Deford's Boordy has expanded the European vinifera varieties. Rob is taking one of these varieties to create something new to the state. "It's Maryland's first champagne. We make it from Chardonnay," says Julie.

A guide will explain the champagne-making process on the thirty-minute tour. You'll walk through the fermentation area, aging room, and bottling line. The tasting could include Chardonnay, Cabernet Sauvignon, Seyval Blanc, Vidal Blanc, and a variety of blends such as Boordy Blush. The wines sell for $5.25 for the Maryland Premium Red and White to $9.50 for the Chardonnay and $11 for the Cabernet Sauvignon.

Tom Burns makes the wine at Boordy. "He and Rob work closely together," says Julie. "Rob found he had to hire someone to do what he originally wanted to do." But Julie says Rob still likes running the winery. "It's really exciting, a fun process."

You may want to give yourself a little extra time to spend at Boordy. You can get a map of the grounds, walk through the vineyards, picnic on the terrace, visit the circa-1776 cemetery on the property, and stroll among the old farm buildings.

The winery also offers numerous special events. In the spring a large open house celebrates the release of the new wines and champagne, and the fall open house marks the release of the Nouveau wine. In the winter there are barrel tastings, and the summer has Sunday cookouts and Maryland country dinners. For the latest dates and times, call the winery.

DIRECTIONS: From 695, the Baltimore Beltway, take exit 29, Cromwell Bridge Road. Turn left and drive north 2.9 miles until the road dead-ends. Turn left onto Glen Arm Road and drive 3.2 miles. Turn left onto Long Green Pike for 2 miles; winery is on the left.

HOURS: Tours and tastings Monday through Sunday on the hour from 1 P.M. until 4 P.M. Winery closed New Year's Day, Easter, Thanksgiving, and Christmas.

EXTRAS: Winery sells wine and wine accessories. Picnic facilities.

Byrd Vineyards & Winery

Church Hill Road
Myersville, MD 21773; (301) 293–1110

William and Sharon Byrd had been warned that the European vinifera wine grapes wouldn't grow in Maryland. But they didn't listen. And the third-oldest winery in Maryland, licensed in 1976, has not only proved the nay-sayers wrong but has also proved that award-winning wine can be made in Maryland.

"Critics have said our Cabernet Sauvignon is the best," says Sharon. And she doesn't mean just critics from Maryland but from all over the United States. "We were the first to win three golds in three consecutive years in the American Wine Competition," says Sharon. Many wineries can produce a gold-medal winner with one year, but it's much more difficult to make consistently gold-winning wine.

Sharon gives the credit to the grapes from the vineyard. "The winemaker can do so much, but if you don't have the grapes . . ." The Byrds' successful vineyard didn't start out as a vineyard. The winery sits atop a hill high above the Catoctin Valley with a panoramic view that is one of the most picturesque in the United States. "We purchased the land before we knew what we were going to do with it. We thought we'd subdivide," says Sharon. But then William checked temperatures and soil conditions and, Sharon says, "It looked like a potentially good spot." So the Byrds jumped into it, beginning with a nursery and selling vines to wineries throughout the East. And finally they produced wine.

But not all has been rosy. "In 1982 we had a cold winter, and in 1985 it was the frost." Cold and unpredictable weather has kept the Byrds from having a crop each year from their thirty-acre vineyard. So in order to have wine to sell when they didn't have a crop the Byrds bought wine from another company and released it under their Byrd Cellars label.

Now the Byrds use only their own grapes. They produce Chardonnay, Sauvignon Blanc, Gewürztraminer, Cabernet Sauvignon, and Riesling, priced from $7.50 to $15.50.

You can see how they're made during the Byrds' fifteen-minute slide show. After the slide presentation you step into the tasting room for samples. The tasting room is in the basement of the red-brick house where the Byrds live, on the land where their vines try to survive.

DIRECTIONS: From Frederick, drive 10 miles west on Interstate 70. Take exit 42, Myersville. Drive north to Church Hill Road. Turn right and drive 1.1 miles; winery is on the left.

HOURS: Slide show and tastings Saturday and Sunday from 1 P.M. until 5 P.M. except during daylight savings time when the hours are 1 P.M. until 6 P.M. Winery closed January, Easter, and Christmas.

EXTRAS: Winery sells wine and wine accessories.

Linganore Winecellars

13601 Glissans Mill Road
Mt. Airy, MD 21771; (301) 831–5889 or (410) 795–6432

"Customers say, 'What's your best wine?' " says Anthony Aellen, winemaker at Linganore. "I say, 'It's the one you like.' " It is with this no-nonsense attitude toward wine that Anthony runs Linganore Winecellars with his brother, Eric, who works in the vineyards, and his parents, Jack and Lucille, who own the winery.

"We don't have the snob appeal that some wineries have. The romance has been overblown," says Anthony. "My judges are the customers. Everyone has their own tastes, and that's why we produce the variety we do. We have a line of thirty." Some of the wines Linganore produces are Seyval, Cayuga, Vidal, Chancellor, and the blush, Tickled Pink.

Anthony also makes fruit wines such as peach, raspberry, blueberry, and strawberry. The Aellens began producing fruit wines when a grower offered them a hundred tons of plums. He decided to give plum wine a try, as he's tried other new wines. "We'll try anything once, and drop the ones that don't work." The fruit wines are produced in two different styles: the sweeter under the Berrywine label and the dry under the Orchard Crest label.

Other wines that have worked for Linganore are spice wines made to be served hot during the holidays and what they call a "conversation leader," Dandylion wine.

Linganore wines are interesting, and so is their pricing system. Prices are keyed to the color of the capsule on the bottle. Prices range from $5 for the black to $9 for the silver caps. Most colored capsules and prices for the varietals are $7 to $8.

The Aellens, with a Swiss-German heritage, have been winemakers for generations. "My husband is a chemist and a home winemaker," says Lucille. "We started with all my father's home equipment." You can see the equipment, housed in a turn-of-the-century dairy barn, on the tour, which is given by a member of the family. "If there's something going on, we'll show visitors," says Lucille. "If not, it can all be seen on our videotape of the process."

After an explanation of winemaking, the Aellens offer a tasting of all their wines. The tasting may run more than half an hour because they're more than happy to explain each wine and why it's special. For example, Lucille will tell you, "We are the only winery in Maryland that makes honey wine. It's made from 100 percent Frederick County honey. It's excellence is attested to by the awards it has won."

If you get Anthony talking about winemaking, he'll show you another side of Linganore. "It's a lot like art. Every year you're given a clear canvas. You can create something different each year," says Anthony. Finally he states what is perhaps the real truth behind the winery: "When you really get down to it, winemaking's more creative than a science."

DIRECTIONS: From Interstate 70, take exit 62, the Mt. Airy exit, or from Interstate 270, take exit 22. Drive north 5 miles on Route 75 to Glissans Mill Road. Turn right, east, and drive 3.7 miles; winery is on the right.

HOURS: Tours and tastings Monday through Friday from 10 A.M. until 5 P.M., Saturday from 10 A.M. until 6 P.M., and Sunday from noon until 6 P.M. Winery closed major holidays.

EXTRAS: Winery sells wine, wine accessories, and gifts.

Montbray Wine Cellars

818 Silver Run Valley
Westminster, MD 21158; (410) 346–7878

"I was raised by a Methodist, a teetotaler," says G. Hamilton Mowbray, owner of Montbray Wine Cellars. So for many years as a young man Hamilton never tasted wine. Then he found himself and his wife, Phyllis, immersed in it. "I was in Europe for a number of years with my wife. I went to Italy and France and found it was a way of life. I got back here in 1953. I couldn't afford the stuff I had been drinking, and California wine was terrible." But what could a research psychologist at Johns Hopkins University do about it?

Hamilton found a book by Philip Wagner, then-owner of Boordy Vineyards in Maryland. Armed with information from Wagner, Hamilton began making wine. In 1964 he went commercial, thus making Montbray the oldest family-owned, continuously operated winery in Maryland.

But Maryland's laws made it difficult, almost impossible, for the small wineries, says Hamilton. "You could be your own wholesaler, but you couldn't sell wine at the winery. At first you could sell one bottle to one person per tour per year." Hamilton says that was a nightmare, trying to keep track of people and if they had already bought their *one* bottle for that year. The laws were finally changed, and now you can buy as many bottles as you like.

Hamilton sells Seyve-Villard for $7, Cabernet Sauvignon for $14, and Cabernet Franc for $18. He also produces a wine from Maréchal Foch called Garnet for $6, and sometimes a Riesling and Chardonnay. "We have about twelve acres of grapes," says Hamilton. His winery capacity is approximately 7,000 gallons. "It's a typical European winery. I think the French make the best wine in the world. All of my wines are in French style." He'll explain how he does it on his hour-long tour.

Hamilton has great credentials for giving winery tours and teaching about wine. He regularly lectures and has written articles on wine appreciation. A patient teacher, he'll answer all your questions. For his work in making wine an easily understood topic, he was awarded the Croix de Chevalier du Mérite Agricole from the French government and the Award of Merit from the American Wine Society.

Montbray forms the perfect setting for sampling wine. Located in a barn on a farm that dates back to the 1700s, the winery has wood-plank floors, huge hand-hewn beams, and oak barrels everywhere.

When you visit the winery in its rustic setting, Hamilton makes the whole operation look easy, but it's not. "We're pretty proud that we do it in traditional European style with all the oak. I work my butt off trying to keep the oak clean," says Hamilton. But his hard labors make for a wonderful afternoon visit.

DIRECTIONS: From Baltimore, drive west on Interstate 40 to the Westminster exit. Drive north 8 miles on Route 97. Turn right on Silver Run Valley Road and drive 1.9 miles; winery is on the left.

HOURS: Tours and tastings by appointment only.

EXTRAS: Winery sells wine.

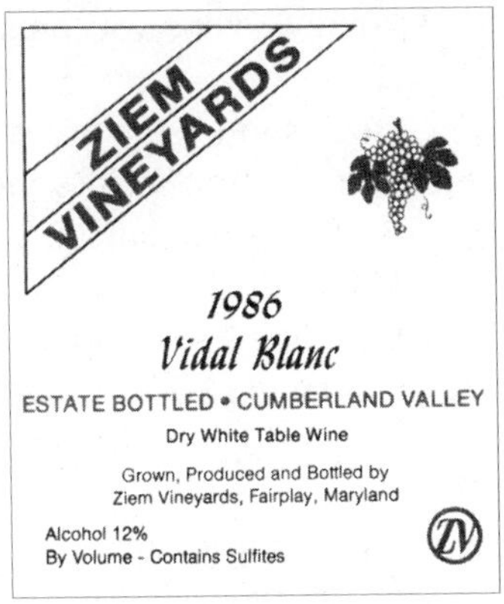

Ziem Vineyards

16651 Spielman Road
Fairplay, MD 21733; (301) 223–8352

Ruth Ziem used to ask herself, "Why would anyone want to wear my name on their shirt?" Then she realized people enjoy being associated with the small winery because of the product and the owners. Ruth and her husband, Robert, run the down-home family winery. Ruth's modesty is surpassed only by Robert's staunch belief in his product.

"I'd like people to know that we're making the best hybrids. You can make fine wines out of hybrids," says Robert. If you mention that some people believe great wines can only be made from the European vinifera, not the French-American hybrids, watch out. "My feeling is, hogwash. Nobody in the East is making better reds than me. If you want vinifera buy French or California wines. We should be offering what we can do best, not simply going for the name recognition wines like Cabernet Sauvignon, Pinot Noir, and Chardonnay."

And he makes a good argument. "Hybrids are synonymous with improving the breed, like roses and corn. The only hope for the Eastern wine industry is the hybrids. The bottom line is, nature has the last word, and she's going to wipe you out [in a cold year]," if you plant vinifera.

Of course, the only proof is to try the wines yourself. The Ziems make only estate-bottled wines. White wines include Vidal, Seyval, and Vignoles. Red wines include Chamborcin, Chancellor, Maréchal Foch, Landot Noir, Léon Millot, Chelois, and DeChaunac. Wines cost $6 to $10. "They're all table wines," says Ruth. "Nothing sweet. And each one gets his personal attention." Robert adds, "If it's not a good wine, I'm not going to bottle it." His wines have won international awards.

From their eight acres of grapes, they produce 2,000 to 2,500 gallons of wine. Ruth says she never knew what they were in for when they started the business. "Twenty years ago we moved up from Washington, D.C. We thought it would be a good retirement business. We never knew how much work it would be." She points to a table filled with glue and labels. "I get to do all this by hand." She says she doesn't work in the vineyard because "I'm too stupid to drive the tractor. He says I'm too smart. If I knew, I'd be out there."

Robert or Ruth will take you "out there" and show you the vineyards and how they use large, blue food-grade plastic barrels and oak barrels to make their Maryland wine. "Why try to copy the French?" asks Ruth. "We're trying to produce a Maryland product." Robert encourages everyone to try the state's wine, even the people who have had bad experiences. "Every state makes some great wine and some bad ones." But if sales are any indication, Ziem Vineyards must be doing something right. "Sales are up, and we have a very high percentage of repeat sales," says Robert.

DIRECTIONS: From Baltimore, take Interstate 70 west to exit 29. Drive south on Route 65 for .9 mile and turn right on Rench Road. Drive till the road ends. Turn left on Route 632 and drive south for 3.7 miles. In Downsville turn left onto Route 63; winery is on the right after .2 mile.

HOURS: Tours and tastings available Thursday through Sunday from 1 P.M. until 6 P.M. Winery closed New Year's Day, Easter, Thanksgiving, and Christmas.

EXTRAS: Winery sells wine and wine accessories.

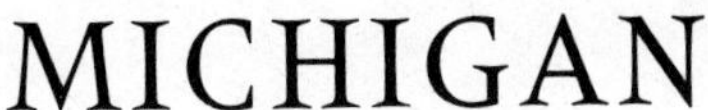

MICHIGAN

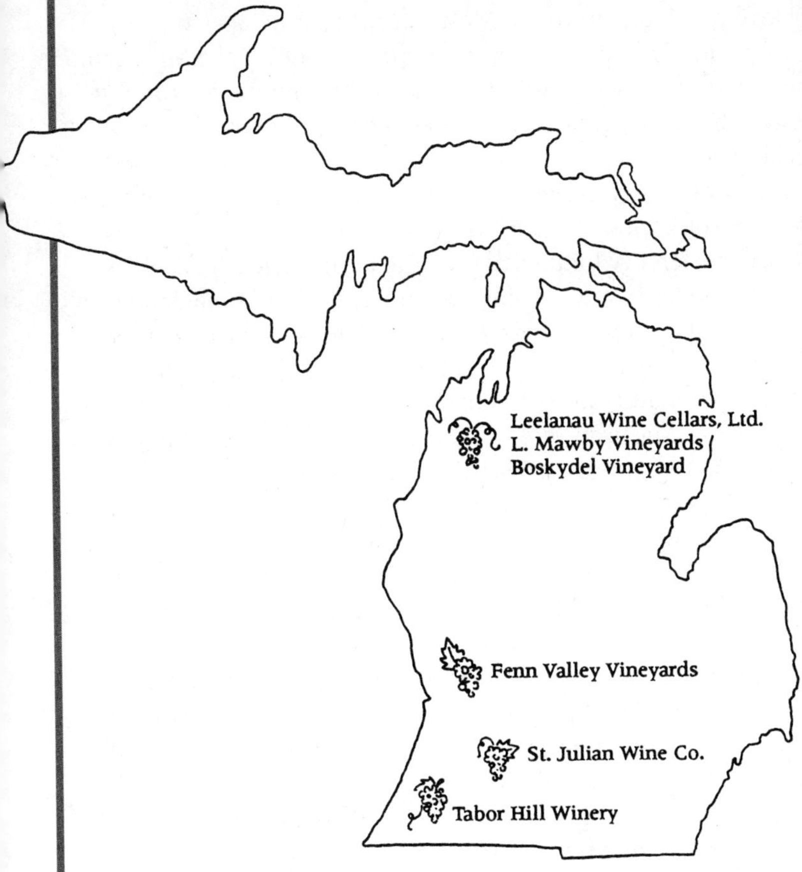
Leelanau Wine Cellars, Ltd.
L. Mawby Vineyards
Boskydel Vineyard
Fenn Valley Vineyards
St. Julian Wine Co.
Tabor Hill Winery

Fenn Valley

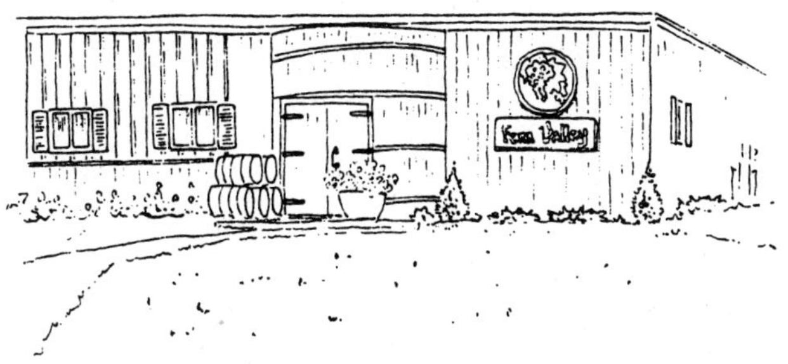

As settlers came to Michigan in the mid-1800s, they brought their love of wine with them. Most of Michigan's grapes were, and are, grown in the southwest corner of the state, near Lake Michigan. According to the Michigan Department of Agriculture, by 1880 the state had more than 2,000 acres of grapes and produced 25,000 cases of wine.

Much of this early wine was made from the native Concord grape. Concord wines are not as popular now due to the public's palate and preference for the flavor of California-style wine.

Today wineries are growing in number, size, and reputation in the state as cultivation of French hybrids and vinifera grapes increases. Growing these nonnative grapes is more difficult. The grapes must be selected for their adaptability to the climate. What makes grape growing possible in Michigan, however, is Lake Michigan and the other bodies of water in the state, which moderate the temperatures.

Lake Michigan absorbs the heat during the summer and slowly releases it in the fall, lengthening the growing season. The cool breezes from the lake in the spring keep the plants from budding too quickly—while there still is a danger of frost. This is the reason why most vineyards are located near the lake.

The state has four appellations—that is, federally defined grape-growing regions that differ from other areas. The largest follows the Lake Michigan shore from Saugatuck in the north to Indiana in the south and as far east as Kalamazoo. The Fennville appellation is to the north of the Lake Michigan shore appellation. An eastern appellation is in the Saginaw and Flint area. A northern appellation contains the Old Mission Peninsula and the Leelanau Peninsula by Traverse City.

Boskydel Vineyard

Route 1, Box 522
Lake Leelanau, MI 49653; (616) 256–7272

If you visit Boskydel when it's not too busy, you're liable to find Bernard C. Rink, the owner, sitting outside his winery's underground cellar, enjoying the beautiful view. The winery overlooks the vineyards and Lake Leelanau. The vineyards rest on a pine-topped slope, facing southwest.

But this doesn't mean Bernard and his wife, Suzanne, are relaxed when it comes to quality wine. Their venture was Leelanau's first bonded winery. The vines, planted in 1964, are French-American hybrids, and the wines produced are Soleil Blanc, Seyval Blanc, Aurora Blanc, Roides Rouge, Rosé du Cru, and Rosé de Chaunac. The wines are priced from $3.35 to $7.25.

You can taste and buy wine year-round. Regular tours aren't offered, but you may talk with Bernard; he has the personality of a fine dry wine—practical but complex.

You may notice the absence of oak barrels. Bernard will tell you that putting oak chips in the wine "does the job, and they're cheaper, too." And while some wineries spend thousands of dollars on netting to protect the grapes from animals, Bernard will tell you, "I don't mind sharing with the birds and the deer."

He also has figured out his own way of protecting his vines from freezing in the winter. Some vineyards use tall pieces of wood put next to the vine to keep the sunshine from the reflecting snow off the grapevine. If left unprotected, the sun will heat the vine. Sap then begins to flow, and when temperatures go below freezing at night, the vine splits and dies. Bernard, on the other hand, simply plants two vines, one to produce the grapes and the other to protect the first.

Be prepared to enjoy a winery that makes wine a "practical way," not necessarily the way it's "supposed" to be.

DIRECTIONS: From Suttons Bay, take Highway 204 toward Lake Leelanau. Turn south on County Road 641. Continue 3.5 miles until you reach the corner of County Road 641 and Otto Road. Turn left on Otto Road; winery is on your left.

HOURS: Tastings are available year-round from 1 P.M. until 6 P.M. daily.

EXTRAS: Winery sells wine.

Fenn Valley Vineyards

6130 122nd Avenue
Fennville, MI 49408; (616) 561–2396

Fenn Valley is located in Michigan's first federally approved viticultural area. The Welsch family, the owners, chose the area for their winery because of the rolling hills, sandy soil, and nearby Lake Michigan, which moderates the temperature. A true cellar, the winery was built into a small hill in the middle of the vineyards.

When entering the visitors area, you walk through a large cask on the front of the building. The Fenn Valley tasting room has three levels, and each level offers the visitor something different.

The first floor of the large room has a horseshoe-shaped tasting bar. A two-part, twelve-minute video is offered in a small room on the second floor. The first section is a seven-minute summary of how wines are made, beginning with the pruning of the vines in the middle of the winter and ending with the final bottling more than a year later. Part two explains how to select and serve wine.

On the third floor is the self-guided tour. From this level you can look out over the winery to see the fermentation, storage, and bottling areas of the winery. Panels on the wall offer information on everything from corks and capsules to bottling and barrel aging. Another panel tells about cold stabilization and why the wineries in Michigan have it better than the ones in California. When Fenn Valley wants to use cooling in the winemaking process in the winter, all it does is let the cold in—not something Napa Valley can do.

If you want to learn still more, there are informational displays on almost all of the walls. One, entitled "Label Wise," explains the terms you'll encounter on a wine label, such as *estate bottled, vintage date,* and *varietals.* Another tells how to evaluate a wine by examining the color, clarity, and aroma. You can read that panel and walk down to the tasting bar on the first level to put the information to use.

Fenn Valley produces estate-bottled award-winning wine from both vinifera and French-American hybrids. Some of the vinifera wines are Chardonnay, Johannisberg Riesling, Pinot Gris, Pinot Noir, and Gewürztraminer. The hybrids are Seyval Blanc, Vignoles, Maréchal Foch, Chancellor, Chambourcin, and Vidal Blanc. The winery makes a dry Lakeshore White, Lakeshore Red, and a Lakeshore Sunset that sell for $5.99 each. Two dry red dinner wines are the 1989 Pinot Noir and Chancellor, which sell for $12.49 and $9.49, respectively. Sweeter wines are Briarwood White, Briarwood Sunset, and Briarwood Red, for $5.99.

Specialty wines from Fenn Valley are raspberry, peach, and mulled wine. For the bubbly lover, there are two champagnes, Brut and Demi-Sec. Both sell for $15.49.

Fenn Valley is one of only a handful of wineries that produces de-alcoholized wines. These wines are fermented and aged in the traditional manner and then de-alcoholized before bottling.

Bonded in 1973, the winery began with a thirty-five-acre vineyard and production of 5,000 gallons. The vineyard has grown to fifty acres, and production has jumped to 50,000 gallons. Not all the original varieties remain in the vineyard, however. Fenn Valley is still experimenting to find the best varieties for the region and the state. Six types of grapes have been pulled up because they did not meet expectations. But plenty of varieties have worked well, and they are all waiting for you in the tasting room.

DIRECTIONS: From Holland, drive south on Interstate 196 to exit

34, the Fennville exit. Turn east on Highway 89 and drive 3.5 miles. Turn right onto Sixty-second Street, south, and drive for 1 mile. At 122nd Avenue turn left, east, and drive .4 mile; winery is on the right.

HOURS: Tours and tastings offered Monday through Saturday from 10 A.M. until 5 P.M. and Sunday from 1 P.M. until 5 P.M. Winery closed New Year's Day, Easter, Thanksgiving, and Christmas.

EXTRAS: Winery sells wine and wine accessories.

Leelanau Wine Cellars, Ltd.

Box 68
12683 County Road 626
Omena, MI 49674; (616) 386–5201

"We've become the largest in Northern Michigan and the second largest in the state," says Bill Skolnik, the winemaker at Leelanau Wine Cellars. Production increased from 5,000 cases of wine in 1987 to 22,000 cases expected in 1992, but size certainly hasn't affected quality. "We've won quite a few gold medals," Bill reports.

Bill says visitors award the tour top honors, too. "People think we give the best tours. They understand a little more about what they're tasting." The thirty-minute tour begins outside with information about the vineyards. Then it's a walk past the crusher and an explanation on how grapes are turned into wine—why most wines are fermented in tanks and why Chardonnay is fermented in barrels. The tour concludes with how wine is processed after fermentation.

Leelanau's wine is made from vinifera and French-American hybrids. "We make eighteen different wines and all are available for tasting," says Bill. On its wine list are fruit wines such as cherry for $4.99, strawberry for $6.99, and peach for $5.99. "The strawberry wine has gone over fantastic," says Bill.

Leelanau buys the fruit for those wines from Michigan growers, but it has approximately forty acres of vineyards that provide its grapes. In the mid-1980s birds wiped out all of Leelanau's Chardonnay. It now uses bird netting to protect the Chardonnay, Riesling, and Pinot Noir. The three types of grapes mature just in time for the birds to feast on as they migrate south.

The gold-medal–winning Tall Ship Chardonnay, for $13.95, is so popular, Bill says it sells out immediately every year. The white Vis a' Vis is a barrel-fermented Chardonnay and Vignoles blend. The red Vis a' Vis has Baco Noir and Pinot Noir. "Those two go together naturally very well," says Bill.

Leelanau also produces a series of wines in accordance with the seasons: Winter White, a semisweet; Spring Splendor, a little drier than the Winter; Summer Sunset, a blush; and Autumn Harvest, a dry red. The four wines sell for $5.95 each. The colorful labels for the series are done by a local artist.

DIRECTIONS: From Traverse City, take Highway 22 north for approximately 25 miles through Suttons Bay to Omena. In Omena, turn left onto County Road 626 for .2 mile; winery is at the top of the hill on the right.

HOURS: Tastings are available daily from noon until 6 P.M. from mid-April until the end of December. Tours available hourly or when a group forms. Winery open only on Saturdays and Sundays during January, February, and March. Closed New Year's Day and Christmas.

EXTRAS: Wine sold at the winery.

L. Mawby Vineyards

4561 South Elm Valley Road
Suttons Bay, MI 49682; (616) 271–3522

If Bernard Rink, owner of Boskydel Vineyard in Lake Leelanau, Michigan, can be compared to a bottle of fine dry red, Lawrence Mawby, of L. Mawby Vineyards, has the enthusiasm and personality of a bottle of champagne. Perhaps it's no coincidence that sparkling wine is also one of his primary interests.

"Sparkling wine makes sense for Michigan. You harvest one month earlier than table wine," says Lawrence. Then the grapes have less chance of having problems with frost. He makes a champagne that's on the sweet end of a brut. "It's a fruity blend."

L. Mawby Vineyards also produces a semidry fruity blend from Seyval and Vignoles called Sandpiper; a dry Beaujolais-style red from Maréchal Foch called M. Foch; and a wine called Turkey Red. He makes a barrel-fermented Chardonnay and two styles of wine from Vignoles. Lawrence uses grapes from his nine-acre vineyard and grapes purchased from other Michigan vineyards. Prices for his wine range from $6 to $14.

Since his first crush in 1978, Lawrence hasn't had any trouble selling his wines. "Demand exceeded what we could produce, so I found [Michigan] growers interested in growing high-quality grapes." Yet he doesn't plan on expanding too much. "I want to make all the wine myself . . . and I don't like machines." He proudly calls his operation a "one-horse winery."

Most of the wine has minimal handling and is barrel fermented. "I like barrels," says Lawrence. He also likes showing people around. The winery building is not arranged to accommodate large groups or children, however, and there are no public restrooms. But Lawrence will pour any of his wines for tasting, and there is plenty of room outside to walk in the vineyards and sit with a picnic.

If you're looking for an unusual handmade basket, such as one made from corn leaves, you'll find it at the winery. Lawrence's friend Peggy Core has her work displayed on one of the winery's walls. Peggy also draws all the labels for the Mawby Vineyards' wine.

You may see Peggy when you visit the winery. "We decided that Peggy and I would be in the tasting room when it's open so there'll always be someone who knows [about Mawby Vineyards]."

And someone who knows can tell you all those wonderful details about grape growing, like how to keep the birds from eating the ripe grapes. "We've used mylar tape, silver on one side and red on the other, and it sparkles when the wind blows. We've tried kites shaped like hawks hung from weather balloons," says Lawrence. But even with all this, in 1985 Lawrence lost all his red grapes to the birds. "You just hope the weather keeps them moving and migrating south."

Not even the birds can dampen Lawrence's enthusiasm about the vineyards, though: "I think winegrowing is as much fun as wine drinking."

DIRECTIONS: From Traverse City, take Highway 22 north toward Suttons Bay. After 8 miles, turn left on Hilltop Road (look for the red barns at the intersection). After .7 mile, turn right onto South Elm Valley Road. Winery is on your right after .2 mile.

HOURS: The winery is open for tastings May 1 through November 1, Thursday, Friday, and Saturday (although Lawrence says Saturday is not a good day) from 1 P.M. until 6 P.M. Tours are available by appointment only and for groups no larger than eight.

EXTRAS: Winery sells wine.

St. Julian Wine Co.

P.O. Box 127
716 South Kalamazoo Street
Paw Paw, MI 49079; (616) 657–5568

Almost anywhere you go in Michigan, you'll find St. Julian. Touted as Michigan's oldest winery, St. Julian has tasting centers in Paw Paw, Frankenmuth, Mackinaw City, Parma, Union Pier, Traverse City, Turkeyville, and Monroe. Two locations, Paw Paw and Frankenmuth, have a winery to provide the backdrop for a tour. All locations, however, offer a video tour showing the grapes from harvest to crushing.

The St. Julian empire began in 1921 in Ontario, Canada. With Prohibition in the United States, Mariano Meconi made his wine across the border from Detroit, calling his winery the Italian Wine Company. After repeal he moved operations first to Detroit and later to Paw Paw to be closer to the grapes. The winery's present name comes from St. Julian, the patron saint of Mariano's birthplace, Faleria, Italy. Although the winery changed its name during World War II, it is still owned by the same family and run by Mariano's grandson, David Braganini.

St. Julian produces almost 150,000 gallons of wine each year.

In 1987 Pope John Paul II selected St. Julian's Seyval Blanc to be used during his mass at the Pontiac Silverdome.

Wines are made from French-American hybrids and vinifera. The white table wines are produced from Seyval Blanc, Vignoles, Vidal Blanc, Chardonnay, Riesling, Aurora, and Cayuga. The red wines include Chancellor Noir, Chambourcin, Foch, DeChaunac, and Chelois. There's also a line of Village wines—Village White, Village Blush, and Village Red—made exclusively from grapes grown in the Lake Michigan Shore appellation. St. Julian offers five sparkling wines, brut, white (extra dry), spumante, raspberry, and cranberry spumante.

And if you still haven't found something you like, there's a full line of table wines as well as specialty and dessert wines. Even if you don't usually like port, you may want to give St. Julian's Solera Cream Sherry a try. This smooth, full-bodied wine has won more than twenty-six gold medals.

For nondrinkers, children, or anyone who enjoys sugar-free juices, there are nine to choose from. There's the Sparkling White Grape Juice and St. Julian's newest, Michigan Cherry Spumante. The premium table wines sell for $4.40 to $9.50, the sparkling wines for $6.25 to $8.50, and the juices for $3.35. The large selection can be overwhelming, but the staff will help you find one, or several, to your tastes. You may choose as many wines as you like to sample.

Before or after your sampling, you may take a St. Julian tour. Offered every hour on the half-hour, the thirty-minute tour will take you winding around old wooden casks and wine aging in oak barrels. The bottling room is lined with windows for you to peek through and watch the latest vintage being bottled. Every machine is labeled so you know what is happening at each stage.

St. Julian is one of the few wineries with a recycling program. You can return its bottles for a five-cent deposit. Many people will find this program to be the perfect reason, or excuse, to keep returning to St. Julian to buy more of its fine wine.

DIRECTIONS: For the Paw Paw location, take exit 60 on Interstate 94. Drive north .5 mile and the winery is on the left.

HOURS: Tastings offered Monday to Saturday from 9 A.M. until 5 P.M., and Sunday from noon to 5 P.M. Tours offered Monday to Saturday from 9:30 A.M. to 4 P.M., and Sunday from 12:30 P.M. to 4 P.M. Winery closed New Year's Day, Easter, Thanksgiving, and Christmas.

EXTRAS: Winery sells wine and wine accessories.

Tabor Hill Winery

185 Mt. Tabor Road
Buchanan, MI 49107; (616) 422–1161

A true year-round winery, Tabor Hill offers hayrides in the spring, summer, and fall and cross-country skiing in the vineyards and rides on a horse-drawn sleigh in the winter. All of this in addition to the opportunity to sample award-winning wine.

Make an appointment for these activities, except for the skiing. Or stop by for any of the scheduled special events: Jazz Fest in July; Harvest Fest at the end of August; and the Neu Wine Fest at the end of September. The festivals provide music, food, grape stomps, and of course, winery tours and wine tasting. Call the winery for specific dates and times. "We do have a good time at our festivals," says Anna Mann of Tabor Hill. "People come back for them every year."

Tours at Tabor Hill begin on the deck outside the winery, overlooking the vine-covered rolling hills and the crusher-stemmer pad. They cover the history of the area and the winery first. "Tabor Hill started in 1970," says Anna. "Four men began the winery. They looked around a long time for a good location—the hills, the soil. It had Concords, so they knew grapes would grow. They changed it to vinifera." In 1981 Dave Upton bought the winery and continues to run it today.

"There were two old barns on the property," says Anna. Look for the wooden beams and the wood from the barns, given new life in the tasting room.

The tour continues in the vineyard among some of the winery's fifty-five acres of Chardonnay, Riesling, Vidal, Seyval Blanc, Vignoles, Pinot Noir, and Baco Noir. "Most of the vineyards are within five miles of the winery. Being close to Lake Michigan is the biggest contributor for us. It's warmer in the winter and cooler in the summer."

At the deep cellar a guide will explain what happens to the grapes. On the vats you'll see seven panels carved by T.C. Cavey, an internationally known artist. You'll also see photographs of some of the famous people who drink Tabor Hill wine: Bob Hope, Burt Reynolds, the Smothers Brothers, and all the presidents at the White House since President Gerald R. Ford.

The tour concludes upstairs with a tasting. Tabor Hill's dry wine includes estate-bottled Lake Michigan Shore Chardonnay, Tabor Hill Chardonnay, White or Red Heritage, and Merlot. For semidry it offers Berrien Blush, Classic Demi-Sec, Riesling, and Gewürztraminer. Sparkling wines include Grand Mark, Grand Mark Rosé, Blanc de Blanc, and Angelo Spinazzé's Spumante. It also makes Hartford Cream Sherry, Late Harvest Riesling, Sangria, and the alcohol-free White Champagne, Raspberry, and Peach. The alcohol-free wines sell for $4.95; the blends for $6.95; and the Grand Mark Rosé sparkling wine, for $21.95.

You may want to come to Tabor Hill ready to taste not only the wine but the food offered in the restaurant as well. Filled with windows offering a view of the vineyard's hills, the restaurant serves up appetizers, salads, sandwiches, and entrees such as raspberry chicken, fresh Norwegian salmon, New England crab cakes, and calico raviolis.

Dinner prices range from $6.25 for a mesquite-grilled shrimp appetizer, to $4.50 for chicken pesto salad, to $16.50 for prime filet mignon and mesquite-grilled shrimp. From December to March lunch and dinner are offered Friday, Saturday, and Sunday. From April to November lunch is offered Wednesday through Sunday and dinner from Wednesday through Saturday. The restaurant is closed December except for parties.

If you have skis or an appetite, bring them to Tabor Hill.

DIRECTIONS: Driving south on Interstate 94, take exit 27, M-63, and drive south toward Niles on Niles Avenue. After .2 mile turn right on Hollywood Road and drive 8.3 miles. Following Tabor Hill's signs, turn left and drive for 1.3 miles; then turn right. Continue for 1 mile and turn left onto Mt. Tabor Road. After .1 mile winery is on the left. Driving north on Interstate 94, take the Bridgman exit, exit 16. Drive 1 mile north to Bridgman and Lake Street intersection. Turn right (east) and follow the signs after driving through town.

HOURS: Tours available from June 1 to August 1, daily from 11:30 A.M. to 4:30 P.M., and from September through December 1 and April through May, on Friday, Saturday, and Sunday from 11:30 A.M. to 4:30 P.M. Tours available other times by appointment. Tastings available daily, year-round, from 10 A.M. to 5 P.M., with tastings available until 8 P.M. in the summer. Winery closed New Year's Day, Thanksgiving, and Christmas.

EXTRAS: Winery sells wine and wine accessories.

MINNESOTA

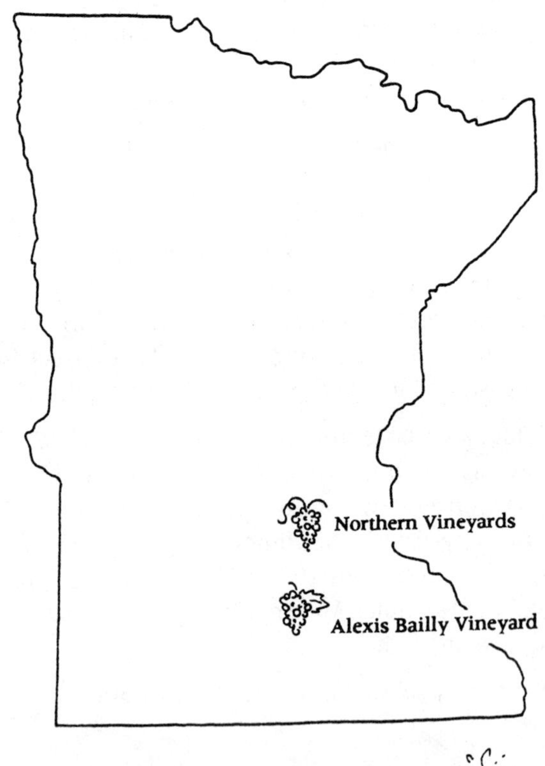

The state that is home to International Falls, a place known for having extremely cold winters, would hardly seem the place to grow grapes. But the Minnesota Grape Growers Association, founded in 1975, sees the weather in Minnesota as no more than a small inconvenience.

Grapes are not completely foreign to the state. The wild grape *Vitis riparia* was always found there. Unfortunately, the wine from the grape is not considered palatable. Vineyard owners are searching for other grapes or crosses with the *riparia* that will survive the cold and have a chance to ripen in the short growing season. "Historically there have never been any wineries, but there was a history of grape growing," says David Bailly of Alexis Bailly Vineyard. Minnesota has had a few wineries through the years, but they made wine from grapes or juice concentrates shipped from out of state.

Minnesota's first winery that made wine with Minnesota grapes opened its doors in 1976. The winery was Lake Sylvia Vineyard and was the first one in the state since the 1800s. The winery's owner, David Macgregor, grew several varieties of grapes at his five-acre vineyard. It has since closed.

Another winery that opened in 1976, and stayed open, is the Alexis Bailly Vineyard, located south of the Twin Cities in Hastings. David Bailly first planted ten acres of vines in 1973. The vineyard has grown to twelve acres, and David has been joined at the winery by his daughter, Nan. David has had luck growing French-American hybrids, and he's still experimenting with other varieties to learn what else can survive the state's temperatures.

Alexis Bailly Vineyard

18200 Kirby Avenue
Hastings, MN 55033; (612) 437–1413

"You are now at the coldest commercial place where they grow grapes," says David Bailly. He and his daughter, Nan, own the vineyard. Why would anyone willingly want to try to grow grapes in Minnesota? "Because I live here," says David with a matter-of-fact tone. "You have to bring it to you or go to it. Plus, the world is full of wineries in California. Any fool can grow grapes in California."

Not just anybody can grow wine grapes in Minnesota. There's a lot of work involved, especially in protecting the vines. "We'll make a trench, lay them to the ground, and then cover the trench. That's the only way we can get them to survive," says David. There's only one variety, which is a cross with the native wild *Vitis riparia,* that the Baillys don't have to bury. And Nan adds, "The grape is one of the three fruits native to Minnesota."

David says the area receives about 50 inches of snow a year, which is good because of the insulation and protection snow offers the vines. "Years we don't have snow we have serious problems. If we could depend on the snow, we wouldn't have to bury as much," says David. "It [Maréchal Foch] survives twenty below real well, but we have thirty below all the time, and sometimes forty below, so we bury it."

After the vines are taken off the trellis and placed on the ground, they're covered with 6 to 10 inches of soil. The Minnesota snow further covers the vines to protect them.

The Bailly vineyard is located approximately thirty minutes from the Twin Cities, in the Upper Mississippi River Valley. "I bought this land in 1972. I was a wine drinker not a winemaker," says David. He planted ten acres of vines in 1973. By 1976 he licensed the winery and became the first to make wine from 100 percent Minnesota-grown grapes. The Baillys grow 95 percent of the grapes for their wine on the twelve-acre vineyard, and buy some from area vineyards. "We grow Maréchal Foch, that's our main grape, the most winter hardy, and Léon Millot, Seyval Blanc, and Minnesota hybrids crossed with wild grapes," says David.

"The reds do very well here, so we'll stick to the reds. We age all our reds in oak. We don't age any of our whites. We make about 5,000 gallons a year in a good year." From the harvest the Baillys produce two whites, a dry Seyval Blanc and a dry Country White. The reds are Maréchal Foch, Léon Millot, and Country Red. Late Harvest, a blend of Riesling and Swenson Red, is a sweet white German-style dessert wine. They also produce a fortified wine called Hastings Reserve. The award-winning wines sell for $5 to $8.

If you like wood, you'll love the tasting and aging room. A large room that takes up the whole bottom floor of the building, it has floors and ceilings made from Minnesota pine. Oak barrels also fill the room, stacked three high. The tasting bar is a board held up by oak barrels. In the tasting room you'll meet either David or Nan. As Nan says, "I'm out here every day." But her father spends Monday through Thursday at his law practice in Minneapolis.

Although there's no formal tour, if the Baillys are working in the vineyards when you visit, they'll let you follow them around and ask questions while they work. And they will explain, as you taste, how each wine was made, blended, and aged. If you have the opportunity to talk with David, he may tell you about the man for whom the winery was named.

"My third great-grandfather [Alexis] came to Minnesota in 1820—quite a controversial person," says David. "He was one of the founders of the local town here. He founded quite a few towns. They kept running him out of town so he had to go to another one."

Mavericks must run in the Bailly family. But David doesn't believe he's doing something off the wall. "Every place has advantages and disadvantages; you just have to take advantage of your advantages." And visitors can take advantage of good wine at the place, as the Bailly label says, "Where the grapes can suffer."

DIRECTIONS: From the Twin Cities drive south for 16 miles on Highway 61 through Hastings. At 170th Street turn right and drive 2 miles (170th turns into Kirby). Winery is on the left.

HOURS: Tastings offered Friday, Saturday, and Sunday from noon until 5 P.M. June through October. Tours available by appointment for groups.

EXTRAS: Winery sells wine.

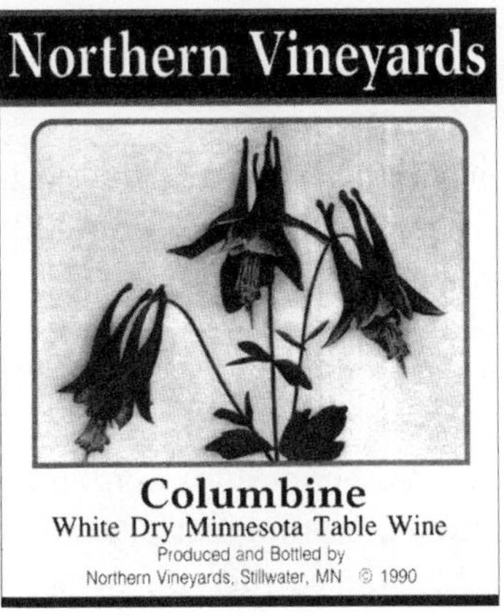

Northern Vineyards

402 North Main Street
Stillwater, MN 55082; (612) 430–1032

Northern Vineyards wines are produced by the Minnesota Winegrowers Cooperative. More than ten vineyard owners, scattered throughout Minnesota and Wisconsin, gather to produce wine at a central winery that was licensed in 1983.

The co-op members realized that with their small vineyards it would be more economical if they united to produce the wine and market it under one label. The vineyards are located in such towns as Shafer, Welch, Rochester, and Redwood Falls in Minnesota, and Hudson and River Falls in Wisconsin. The co-op has more than thirty acres of vines. It produces approximately 5,000 gallons of wine each year. Much of the equipment the co-op uses is from the now-defunct Lake Sylvia winery. Its previous owner, David Macgregor, is a member of the co-op.

The Yellow Moccasin, a semisweet white, is made from Aurore and Edelweiss grapes. The Lady Slipper, a blush, is produced from white grapes with a little red St. Croix mixed in for color. The Columbine has a blend of Elvira and Aurore. The dry red St. Croix is made from a local grape variety and barrel aged for a year. There's also a Maréchal Foch Rosé with a bright cherry-pink color. The wines sell for $4.50 to $7.95.

The winery and tasting room are located in the historic Isaac Staples Mill on the northern edge of Stillwater. You can see pictures of vineyards and grapes on the walls. At the back of the tasting room is a large glass window where you can look into the winery. Unfortunately, most of the view is of cardboard boxes. If you want to see the winery in operation, you must make an appointment.

But there's plenty to see in Stillwater, known as the best place for antiques in Minnesota and a great place to see the fall color in September and October. Called the "Birthplace of Minnesota," the town was founded in 1839. Many of the historic buildings have been renovated and are open for visitors. Other old buildings have been converted into shops. The mouth of the St. Croix River lies just north of town, and at Stillwater it widens to form a lake. This natural boundary between the shores of Wisconsin and Minnesota is filled in the summer with sailboats, waterskiers, and canoeists.

Lowell Park sits right on the St. Croix and provides the perfect setting for picnicking and drinking wine. From the park you'll be able to watch the unusual drawbridge raise to let boaters through. Unlike most bridges where one end of the bridge rises vertically, one section of this bridge lifts horizontally at both ends. If you're looking for more than a picnic in the park, the Minnesota Zephyr offers dinner while it whisks you along railroad tracks that follow the St. Croix River.

DIRECTIONS: From the Twin Cities, drive east on Highway 36 to Stillwater. Drive north on Main Street (Highway 95); winery is on the left.

HOURS: Tastings available Monday through Saturday from 10 A.M. until 5 P.M. and Sunday from noon until 5 P.M. Tours by appointment during hours winery is open for tasting. Winery closed New Year's Day, Easter, July 4, Thanksgiving, and Christmas.

EXTRAS: Winery sells wine and wine accessories.

MISSISSIPPI

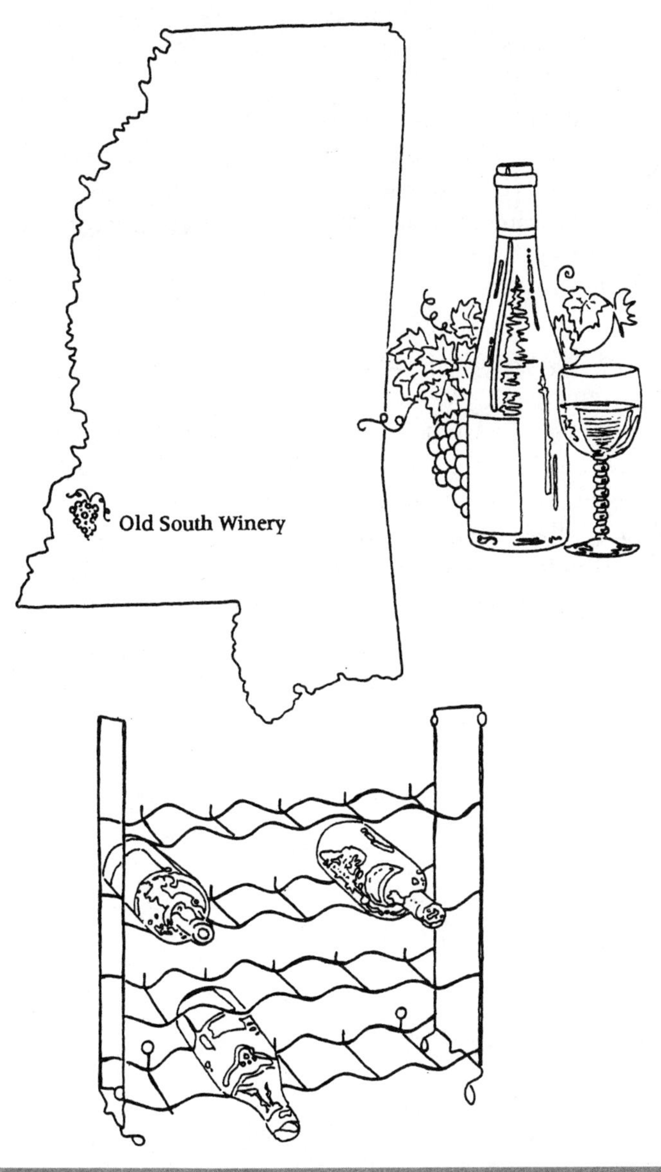

The native American Muscadine grapes have been used for winemaking in Mississippi since the French Huguenots settled in the area in the mid-1500s. Winemaking continued in the region until Prohibition. It has taken almost fifty years for the state's winemakers to recover.

The Winery Rushing received Bonded Winery Permit Number One in 1977, making it the first winery in Mississippi since the days of Prohibition. (Unfortunately, it went out of business in 1990.) The next leap forward for winemaking in Mississippi came in 1984 when the federal government named the Mississippi Delta a designated viticultural area, or appellation. An appellation is a grape-growing region whose geographic and climatic features set it apart from other areas. At 6,000 square miles it is the second-largest appellation in the country.

Although the word *delta* usually refers to the area by the mouth of a river, the Mississippi Delta appellation contains the region several hundred miles north of the current mouth. When its mouth was farther north, the river left rich deposits. These deposits make fertile soil with topsoil often 35 feet deep. The grape growers in the area believe the soils will in the future help to make Mississippi known for its wine.

Old South Winery

65 South Concord Street
Natchez, MS 39120; (601) 445–9924

The Old South. The winery's name says it all—tradition, pride, grandeur. According to Scott and Edeen Galbreath, the Galbreath ancestors brought Muscadine vines from South Carolina to Mississippi in the early 1800s—tradition. "We try to make a little better wine," says Edeen—pride. The Old South labels feature scenes of antebellum Mississippi with sketches of plantations and Southern belles, and the wines have names such as Miss Scarlett, Magnolia, and Southern Belle. One wall in the main room of the winery is covered almost from floor to ceiling with a colorful mural of the *Delta Queen* paddleboat—grandeur.

Edeen begins the thirty-minute tour of the winery in a small room, like a sitting room, full of old stuffed chairs and a couch. The walls are decorated with color photographs of Muscadines, a clock that says "It's time for wine," and a needlepoint map of Mississippi that Edeen's daughter made. Edeen points out the specialty of each area of the state, such as soybeans or catfish, as shown in the needlepoint map. Then you walk through the winery and Edeen explains the winemaking process.

The tasting includes any or all of the Muscadine wines: Miss Carlos, Miss Carlos Dry, Carlos, Carlos Dry, Sweet Noble, Noble, Dry Noble, Noble Rosé, Sweet Magnolia, Southern Belle, Miss Scarlett, and the blush Back to America. The Carloses are white wines and the Nobles are red wines.

All the wine labels commemorate an element of the Old South that has been preserved and may be enjoyed today. Back to America celebrates the Natchez Trace with a picture of a horse-drawn wagon with the sign "Natchez or Bust." Begun as an Indian path, the Natchez Trace has been found on French maps of the area as early as 1733. Crossing the state from northeast to southwest, the Trace became a frontier trail between Washington, D.C., and the Mississippi territory. Today the Trace is a parkway for cars. Driving along the parkway you can still see imprints from the original Trace. The parkway features roadside exhibits, camping, and incredible scenery—a great place for a picnic with a bottle of Back to America.

The Magnolia wine has a sketch of Magnolia Hall. This plantation, which Edeen recommends you visit while you're in the area, is considered to be the last great mansion constructed in the area before the Civil War. Built in the Greek Revival style, Magnolia Hall is filled with antique furniture and paintings from the 1700s and 1800s.

The Miss Scarlett wine features a drawing of Longwood, built by the rich cotton grower Dr. Haller Nutt. The octagonal house was to have thirty-two rooms, each with its own entry or balcony. The fifth floor was to be used as a solarium and the sixth as an observatory. The Civil War began during its construction, and only the cellar and first floor were completed. Nail kegs and paint cans were left on the upper floors as the workers, mostly Northerners, left to join the fighting.

Sweet Noble has a picture of Connelly's Tavern. Built in the late 1790s and restored and authentically furnished in 1935, Connelly's Tavern is considered one of the must-see buildings in the area. Pat Connelly's inn and tavern was touted as the last stop in civilization when the Mississippi was America's gateway to the Southwest. For fourpence a night you had a bed, or for sixpence supper was included. The tavern rules reflected the times: "No more than five to sleep in one bed." "No boots to be worn in bed." "Organ Grinders to sleep in the wash house."

And finally, Miss Carlos carries a drawing of Stanton Hall—a white mansion that rests under giant oaks. Begun in 1851, it took Frederick Stanton six years to build. Wanting only the best and having the money to do it, Stanton sent his architect, Thomas Rose, to Europe to buy furnishings: French mirrors with gold leaf, Italian-carved marble mantels, British silverwork. Stanton enjoyed his new home for less than three months before he died.

Old South's wine is as enjoyable as its labels and the local sites. The Galbreaths do a commendable job of extracting various tastes from Muscadines, a grape known for its usually strong flavor. The medals displayed behind the tasting counter attest to that fact.

While you're tasting wine, Edeen is likely to tell you how good it is for you. "It has iron you can absorb, and it takes cholesterol out. It's nature's tranquilizer, and it's not the alcohol that does that. It takes three to four ounces, and you need to drink it all at once. It gets those cobwebs out."

Every wine sells for $5.50 or $5 each when you purchase three or four bottles. For that price you can't afford not to get healthy.

DIRECTIONS: From Natchez, drive north on Business 61 to Concord Street. (Concord is 1.3 miles south of Highway 61.) Turn west on Concord; winery is .4 mile on the left.

HOURS: Tours and tastings on demand daily from 10 A.M. until 6 P.M. Closed Christmas.

EXTRAS: Winery sells wine, wine accessories, and T-shirts.

MISSOURI

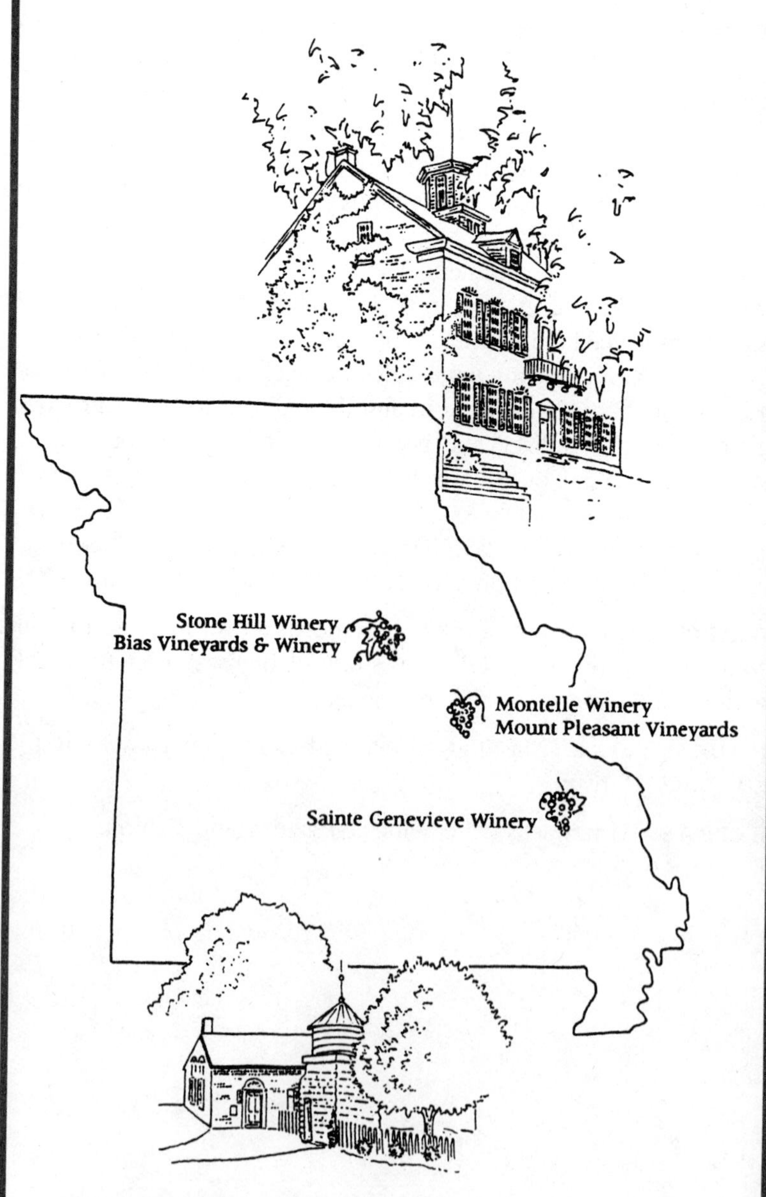
Stone Hill Winery
Bias Vineyards & Winery
Montelle Winery
Mount Pleasant Vineyards
Sainte Genevieve Winery

Missouri wineries date back to 1843 when priests used native grapes to make sacramental wine. German immigrants who came to the area also brought their skills as winemakers with them. Using local grape varieties, vineyards sprouted up all over the state. Soon they were not only plentiful but also large. Stone Hill Winery in Hermann, founded in 1847, became the second-largest winery in the country.

By the mid-1860s the state was producing almost 3 million gallons of wine and was second only to New York in wine production. Toward the end of the nineteenth century, Italians joined the Germans in their winemaking efforts.

Hermann Jaeger, a Missouri enologist, was awarded the French Legion of Honor for his work in saving the great vineyards of Europe. He found his native vines were resistant to the phylloxera root louse that was destroying European vines. In the 1870s Hermann sent his Missouri rootstock to France, where the French grafted their vines onto his. Some of the vines from that original rootstock eventually made their way back across the Atlantic and were used in California vineyards to protect them from the pest.

Prohibition put an end to Missouri's great wine days. Stone Hill Winery turned its wine cellars into mushroom beds, but most other wineries shut down their operations and tore up their vineyards. Even with repeal, some Missouri counties remained dry, and most wineries remained closed. Those that did reopen in the following years found that America's taste in wine had changed, and their sweet wines were no longer in demand as they were in the 1800s.

The renewed interest in American wines during the 1960s and 1970s also revitalized Missouri. Vineyard owners began planting French-American hybrids and once again made wine. Missouri vineyard owners have received help from the state's Department of Agriculture. In 1980 it began a program to study grape varieties and to promote the development of the grape-growing industry.

The Missouri wine industry received another big boost in 1980, when the federal government asked for petitions for appellations, or wine-growing regions. Augusta, Missouri, home of the state's wine industry more than a hundred years before, petitioned and received the first appellation in the United States given by the federal government, designating it as an official wine-growing area.

Bias Vineyards & Winery

Route 1, Box 93
Berger, MO 63014; (314) 834–5475

"We got twenty tons hanging out there," says tour guide Linda Sitton about the seven-acre vineyard at Bias. "You're welcome to take a golf cart and tour the vineyards, except during the festivals, and then we run the tram." If you're interested in touring a vineyard, at Bias you do it with style.

"We have a lot of people come out just to see the vineyards. Some wineries won't tell you where their vineyards are," says Linda. "We also let our customers prune, pick—whatever they want to do. We just let them jump in." If the grapes are ripe, Linda will send you out into the vineyard to taste some, with a warning about the bees that hide behind the bunches.

In the vineyard you can taste Catawba, DeChaunac, Vidal, and

Seyval. Harvest occurs from mid-August to mid-September, so that's the time to visit to taste the grapes. If you're interested in tasting the wine, you may try the estate-bottled River Bluff Rouge, a semidry rosé; Berger Red, a sweet red; River View White, a dry white; or Pink Catawba, a sweet rosé.

Another Bias wine is the Frosty Meadow White, an ice wine. "It's actually made from frozen grapes. It's our most expensive because the grapes are left hanging, and you could lose the whole works" if the weather doesn't cooperate, says Linda. The ice wine, which is a sweet table wine, sells for $8. The other wines sell from $4.50 to $7.50.

You can enjoy a bottle in the Weingarden or sit down by the Biases' pond. The Weingarden is a shady spot in front of the tasting room. There are tables and chairs, an old wine press, and a sundial. "There's nearly always a breeze out there," says Linda. That can make a difference on a hot and muggy summer day in Missouri.

Or if you prefer sitting by water, picnic tables are scattered around the pond down the hill from the winery. Bring some extra food. "You can feed the fish," says Linda.

Through a door in the tasting room you step into the large room that holds the 5,000-gallon-capacity winery. You won't find old oak barrels here. "We use all stainless steel. The smaller amounts just go into beer kegs [for fermenting], and we use it for blending," says Linda. Linda, who has worked at the winery for more than six years, will give you a quick tour around the kegs, or she will explain the winery's operations in depth—it's up to you. Linda will do everything she can to guarantee you have a good time at Bias. "I love the people, and I love the tourists." And it shows.

DIRECTIONS: From Hermann, drive east on Highway 100 for 7 miles. At State Route B turn left and drive to Bias Road. Turn right; winery is at the end of the road. You have to cross railroad tracks on Bias Road, and trains frequently use the tracks. Cross with caution.

HOURS: Tours and tastings offered Monday through Saturday from 10 A.M. until 6 P.M. and Sunday from noon to 6 P.M. Winery closed New Year's Day, Easter, Thanksgiving, and Christmas.

EXTRAS: Winery sells wine and wine accessories. Picnic facilities.

Montelle Winery

P.O. Box 147
Augusta, MO 63332; (314) 228–4464

Winding your way up the hill on a gravel road, you thread your way between trees and vines. At the top you'll find Montelle waiting. "It means literally *little mountain,*" says Robert Slifer. Robert owns Montelle with Judith Slifer and Joanne and Bill Fitch.

Montelle Winery at Osage Ridge is actually two wineries that became one. Montelle opened in 1976 in a different location. "We merged with the Osage Ridge Winery," says Robert. "They had the property, and we had the equipment." So in 1987 Robert moved his Montelle equipment up to Osage Ridge.

You can enjoy a view of the Missouri River 400 feet below while drinking wine at the tables outside. One of Montelle's staff will serve you wine, and you can nibble on the locally made cheese and sausage that are for sale. You'll have more than thirty wines to choose from, such as the dry red Cynthiana and Chancellor or the sweet dessert wines Sweet Briar and Augustaner. Some of the award-winning wines are dry white Vignoles, dry white Vidal, and semidry Autumn Blush. Prices range from $4.99 for the dessert wine Old Fashioned to $17.50 for a dry Cynthiana. The winery offers a full line of wines in various styles, but it "started out with the red wines in the French style," explains Judith Slifer, who's in charge of the tasting room.

While you're sampling wine Judith will explain how each blend is made and offer tips on how to evaluate a wine. For nondrinkers Judith will pour grape juice. If you're interested in how Montelle makes its wines, she'll also take you on a twenty-minute tour.

The winery is practically busting at the seams with stainless-steel tanks, oak barrels, and plastic barrels. You can believe it when Robert says, "We've been in the business for fifteen years, and we've been growing ever since." Judith enjoys pointing out and explaining the role of the fermentation tanks. Montelle ferments its white wine in cooled tanks. "We can ferment at fifty degrees, and that's important, as it improves the flavor of the wine," says Judith.

The winery hosts several special events. In May there's the Spring Wine Fest, which highlights new wine releases and offers catered food to accompany them. Fall wine releases are celebrated at the annual Red Wine Tastefest held in October. And on Saturday nights in the summer, Montelle has Dinners On the Ridge—three-course meals served on the deck. Dinner menus have included lemon walnut chicken, grilled swordfish, and beef in red wine. Call the winery for dates and prices.

The wine you taste may have grapes from Montelle's vineyard or from thirty acres that are under contract with area growers. "We try to get people not to compare these wines with California wines," says Judith. "It's like comparing apples and oranges." But that doesn't mean Judith thinks California wines have Missouri wines beat. "In some respects I think our wines have more flavor." But Judith wants you to be the judge. After all, she's from the "Show Me State."

DIRECTIONS: From U.S. 40, drive southwest on Highway 94 toward Augusta for 15 miles; winery is on the right.

HOURS: Tours and tastings offered Monday through Saturday from 10 A.M. until 5:30 P.M. and Sunday from noon until 5:30 P.M. Winery closed New Year's Day, Easter, Thanksgiving, and Christmas.

EXTRAS: Winery sells wine, food, and wine accessories. Handicapped facilities. Picnic facilities.

Mount Pleasant Vineyards

5634 High Street
Augusta, MO 63332; (314) 228–4419

You can almost feel the weight of the 16 feet of dirt above your head. The wine cellars of Mount Pleasant, built in 1881, were part of the original Mount Pleasant owned by Friedrich Munch, a man famous for his wines and for breeding grapes.

The wine cellar was one of the first to have a curved, vaulted ceiling. The ceiling is made from handmade bricks that were fired behind the winery building. The 3-foot-thick limestone walls keep the temperatures fairly constant. In the summer the temperature rises only to about sixty-five degrees and in the winter drops to forty-five degrees.

The original dirt floors remained until approximately ten years ago when cement was poured for easy cleaning. But the original fireplace is still in place. It's believed the fireplace, which takes up almost one whole wall at the end of the cellar, was used for light. The cellar master didn't want to use oil lanterns because of the impurities that they would put into the air.

When Munch ran Mount Pleasant, there were ten other wineries in the Augusta area. The wineries were winning national and international awards for their wines until Prohibition closed them down. Munch ripped up his vineyard and broke up his wine vats.

Mount Pleasant was reopened in 1968 by Lucian and Eva Dressel. They replanted the original vineyard site with French hybrids and native varieties. The vineyard has expanded from five acres to sixty, and production has grown from 750 gallons a year to 30,000. Mount Pleasant now has the largest commercial vinifera vineyard in the state, with more than 10,000 Chardonnay, Cabernet Sauvignon, Merlot, and Pinot Noir vines.

Some of the wines are White and Red La Rustica for $5.49; Emigré Blanc, Missouri Rhineland, and Blush Nouveau for $6.49; Seyval Blanc and Rayon d'Or for $7.49; Vidal Blanc for $7.99; Brut Reserve for $10.99; and Pearl of Oman Port for $18.98. The Dressels brought back a touch of Missouri's former glory when their 1986 Port won a medal in 1989 in London. Mount Pleasant thus became the first Missouri winery to win an international award in more than one hundred years.

The fifteen-minute tour, which costs $1.50, begins at the vineyards behind the tasting room. The guide explains the winery's history and offers information on what the Dressels are doing today. You'll enter the winery and receive a brief explanation of the winemaking process. The tour concludes in Munch's original cellars.

The glass-enclosed terrace has seating for 300, a fireplace, and a two-level deck overlooking the Missouri River Valley. If you want locally made cheese, sausage, and crackers, Eva's Cheese Wedge, located behind the winery, will provide the food for your afternoon break from touring wineries.

DIRECTIONS: From U.S. Highway 40, take State Highway 94 southwest for 16.4 miles. Turn left onto Church Road to the town of Augusta. Follow signs through town to the winery.

HOURS: Tours and tastings from Monday through Saturday from 10 A.M. until 5:30 P.M. and Sunday from noon until 5:30 P.M. Winery closed New Year's Day, Thanksgiving, and Christmas.

EXTRAS: Winery sells wine and wine accessories. Picnic facilities.

Sainte Genevieve Winery

245 Merchant Street
Ste. Genevieve, MO 63670; (314) 883–2800 or (314) 483–2012

"People are unaware of Missouri wines, but as soon as they taste them, it's a newfound treasure," says Hope Hoffmeister, who owns the Sainte Genevieve Winery with her husband, Linus. "We've always made wine, for as long as I can remember. People commented and said, 'You make pretty good wine,' so we decided to do it commercially."

Hope and Linus Hoffmeister make 5,000 gallons of wine a year at the family home west of Ste. Genevieve. But the tasting room, in what was a private home, is located in the village of Ste. Genevieve. The town, established between 1725 and 1750 by the French, was the first permanent settlement west of the Mississippi. You may still see the French Colonial vertical log homes built by fur traders and farmers. From the tasting room you may walk around the downtown area, which has been designated a historic district.

Although the town is old, the winery is fairly young. Licensed in 1984, the business is beginning to take off. As Hope walks you around, she explains what the Hoffmeisters will do in the future. "This year's project is to put a wine garden between the two buildings," she says as she indicates the large grassy area between the house and the barn.

Their goal was to start small and increase according to the market. Each year the winery has been in business, it has produced only an additional 1,000 gallons, so they're taking it slowly. "A minimum amount that we are going to need is 7,800 gallons this year. That's what our target is. Business is great." Hope says it didn't hurt business any when their Concord won a bronze medal at the 1990 Missouri State Fair or when the dry red Beauvais won a silver medal in 1991.

Other wines are French hybrids such as Seyval Blanc and Vidal Blanc. They also produce blends such as the white dessert wine Amoreaux, the semidry, two-month-oak-aged Ste. Gemme, and the sweet white Valle Rhine. Most wines sell for $6.50 or $7. Fruit wines such as blackberry, red raspberry, apple, plum, elderberry, and cherry sell for $7. "I talk to people and find out what they prefer," says Hope, and then she pours accordingly.

If you're really interested in touring the winery, arrangements may be made to meet someone out at the family farm. But if you have already toured a winery and received a description of the winemaking process, you might prefer to taste the wine and walk around historic Ste. Genevieve.

"We have some really neat bed-and-breakfasts and festivals," says Hope. "The second full weekend in August we have the largest crafts fair in the state. The first weekend in October is German days with dancing and German-style bands. In November we have French Colonial Days with lantern tours, and in December they open up the old historic homes for tours during the Old Christmas Walk.

"We're getting to be *the* place for getaways."

DIRECTIONS: From St. Louis, drive approximately 40 miles south to the Ste. Genevieve exit, the Highway 32 exit. Drive east on Highway 32 for 6.8 miles to Ste. Genevieve. At Merchant Street, winery is two blocks east on the left side of the road.

HOURS: Tours and tastings offered March through December, Monday through Saturday from 11 A.M. until 4 P.M. and Sunday from noon to 4 P.M. In January and February winery open Saturday from 11 A.M. to 4 P.M. and Sunday from noon to 4 P.M. Winery closed Thanksgiving, Christmas Eve, and Christmas.

EXTRAS: Winery sells wine and wine accessories.

Stone Hill Winery

Route 1, Box 26
Hermann, MO 65041; (314) 486–2129

On a hilltop overlooking the town of Hermann rests Stone Hill Winery, with what are said to be the largest series of arched cellars in the United States. Larger than the vaulted cellars at Missouri's Mount Pleasant, this series of cellars took more than twenty years to dig. One is called the Apostles Cellar because of its twelve stone arches.

Now listed as a national historical site, the winery was established in 1847 and at one point in the late 1800s was the second-largest winery in the United States, producing 1,250,000 gallons. When Prohibition closed the winery, the cellars were used to grow mushrooms.

L. James and Betty Held reopened the winery in 1965. They have also opened "Vintage 1847," a restaurant in the original carriage house and horse barn. Located next to the winery, the restaurant specializes in German food such as smoked German sausage and schnitzel. The dark-wood–paneled restaurant also offers such fare as roast rack of lamb with pesto and fresh mint sauce for $13.75 or stuffed rainbow trout Missouriana for $12.50. The menu doubles as a cookbook, showing how each entree is prepared. Hours vary for lunch and dinner, so you'll need to call first.

In the red-brick main house you may sample the fruit of the Helds' labor. Stone Hill produces seventeen different types of wine. You'll be asked if you prefer dry, semidry, or sweet wines. You may want to try at least a few drys before moving to the sweets, even if you normally like sweeter wines. The dry side features Norton, Vidal, Seyval, and a brut champagne. The semidry wines are Hermannsberger, Missouri Riesling, Hermanner Steinberg, and Blush. The sweet side has such wines as Rosé Montaigne, Golden Rhine, Vignoles, Spumante Blush, Golden Spumante, Pink Catawba, Concord, and Harvest Peach. Nondrinkers and children may sample the Catawba grape juice. Wines sell for $5.49 to $14.99.

You may buy your tour ticket in the tasting room. The winery charges $1.50 for adults and 50 cents for children for the thirty-minute tour. The tour guide will tell you about historic Hermann and the Germans that settled it. And you'll walk through the cellars while receiving a description on how they make wine at Stone Hill.

Nowadays you won't see any mushrooms growing in the cellars, you'll just see the return of the tradition that began in 1847.

DIRECTIONS: From Interstate 70, take the Highway 19 exit and drive south for 14 miles to and through the town of Hermann. Turn right on Highway 100 for 2 blocks and turn right on Washington Street. Drive 2 blocks and turn left on Twelfth Street. After one block winery is on the left.

HOURS: Tours and tastings available Monday through Saturday from 8:30 A.M. to dusk and Sunday from noon to 6 P.M. Winery closed New Year's Day, Thanksgiving, and Christmas.

EXTRAS: Winery sells wine and wine accessories.

MONTANA

Mission Mountain Winery

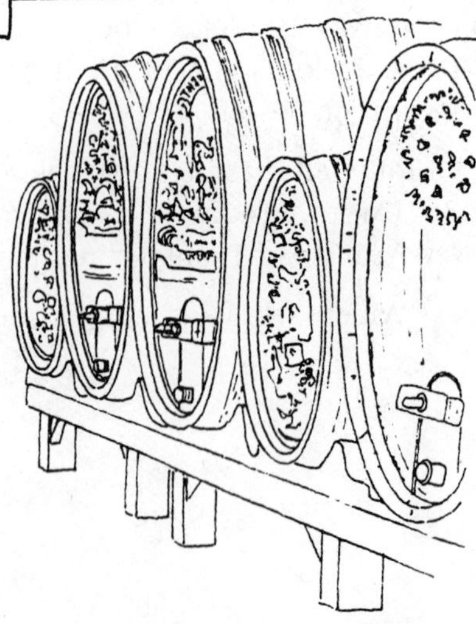

Could Big Sky Country become Big Wine Country? It will be a few years until we know. The first bonded winery in the state has been open only since 1985. But Montana's only winery, nestled in the northern Rockies, appears to have a bright future. If other fruits grow in an area, usually wine grapes will, too; this area is known for its bing cherries.

"We know that around Flathead Lake commercial cherries do very well," says Thomas Campbell, father of Tom Campbell, owner of Mission Mountain Winery. "We tried to find the area that got the most sun, and we think Dayton gets the most." The small town of Dayton is located on the shores of Flathead Lake. "The lake keeps it moderate. We irrigate right out of the lake," says Thomas.

The lack of water isn't the problem. "The main problem is the cold weather, trying to find the plant that will survive the winter. We get about 9 to 10 inches of rain, about 10 inches of snow," says Thomas.

To fight Montana's winters the Campbells have to use special measures. "[The harvest] is usually the last week in September. We bury our vines every fall after harvest. We use topsoil [to cover them], about 6 to 8 inches over the top, and then dig them up in the spring and put them back on the trellises. Usually dig them up around April," says Thomas. "The seven-year-old ones survived not being buried."

Of course the best way to fight the cold is to find varieties that can survive it. The Campbells believe they have found the grape for Montana: the Pinot Noir. "We haven't had to use any chemicals at all," says Thomas.

The Campbells say others have started vineyards in the area, and more are considering tearing out their cherries and planting vineyards. Only time will tell if more vineyards will bring more wineries. Much will depend on the success of the Campbell family.

Mission Mountain Winery

P.O. Box 185
Dayton, MT 59914; (406) 849–5524

"Everything we do is fun because it's the first," says Cheryl Tassemeyer, the manager of Mission Mountain's tasting room. It could be that everything seems fun because the tasting room where she works has such a glorious view of Flathead Lake. "The lake is 35 miles long. In the morning you'll have fifty Canada geese and two blue heron. We have osprey diving out here all the time," says Thomas Campbell, father of Tom Campbell, the owner of the winery.

Even the winery's label, drawn by a Montana artist, features Flathead Lake with sailboats skimming the water in the foreground and snow-capped mountains in the background.

Don't miss the photo album on the tasting bar that shows the winery's grapes, press, and stemmer-crusher. You'll have the opportunity to taste Chardonnay; Johannisberg Riesling; Muscat Canelli, with a sweet start and a dry finish; and Sundown, a medium semisweet blush. The Chardonnay sells for $9.95; the Johannisberg, for $5.49; Sundown, for $5.50; and 1986 champagne, for $19.95.

The Pale Ruby champagne is 95 percent Pinot Noir grown in Montana. You can also buy a bottle of the 1985 champagne for $35.

"It's Montana's first," says Cheryl. Thomas adds, "That is 100 percent Montana grapes off of our vineyard and Finley Point. We only sell it at the winery. [It's] terrific champagne.

"His wines have won national recognition," says Thomas of his son. Tom was born in Missoula and graduated from the University of Montana. He studied at the University of California at Davis, a university famous for its work in viticulture and enology. After college Tom worked in several wineries in Washington. "I always had an interest in winemaking, and my son studied it," says Thomas. "We were in a number of wine-tasting groups. He has a very good palate. He can tell how good the wine will be in two to three years. That's real important for a winemaker."

Tom finally founded his own winery in Zillah, Washington, called Horizon's Edge. With the Washington winery under way, Tom came back to his native state and started planting vines. He began the vineyard in 1980 with five acres. By 1988 he had expanded to eighteen. "This is just starting to come into production," says Thomas, pointing out that vines take five or six years before they reach maturity.

Not everything you taste is 100 percent Montana grown. "The long-range goal is to use Montana grapes. A lot are still coming from Washington, of course," says Cheryl. Thomas adds, "We make a very good Riesling from Washington grapes, so we will continue to import them. [But the] Pinot and also the champagne is from here.

"What we're interested in is producing a quality product. We think we have the chance to make the best Pinot. We already have some Pinot in the barrel," says Thomas. You can tell by the look on his face that he can't wait for it to finish aging so it will be ready to drink.

DIRECTIONS: From Missoula, drive north on Highway 93 about 80 miles. Winery is on the right 3 miles north of Elmo, just before entering Dayton.

HOURS: Tastings and tours available from the first of May until the end of October, daily from 10 A.M. until 5 P.M.

EXTRAS: Wine and wine accessories sold at the winery.

NEW JERSEY

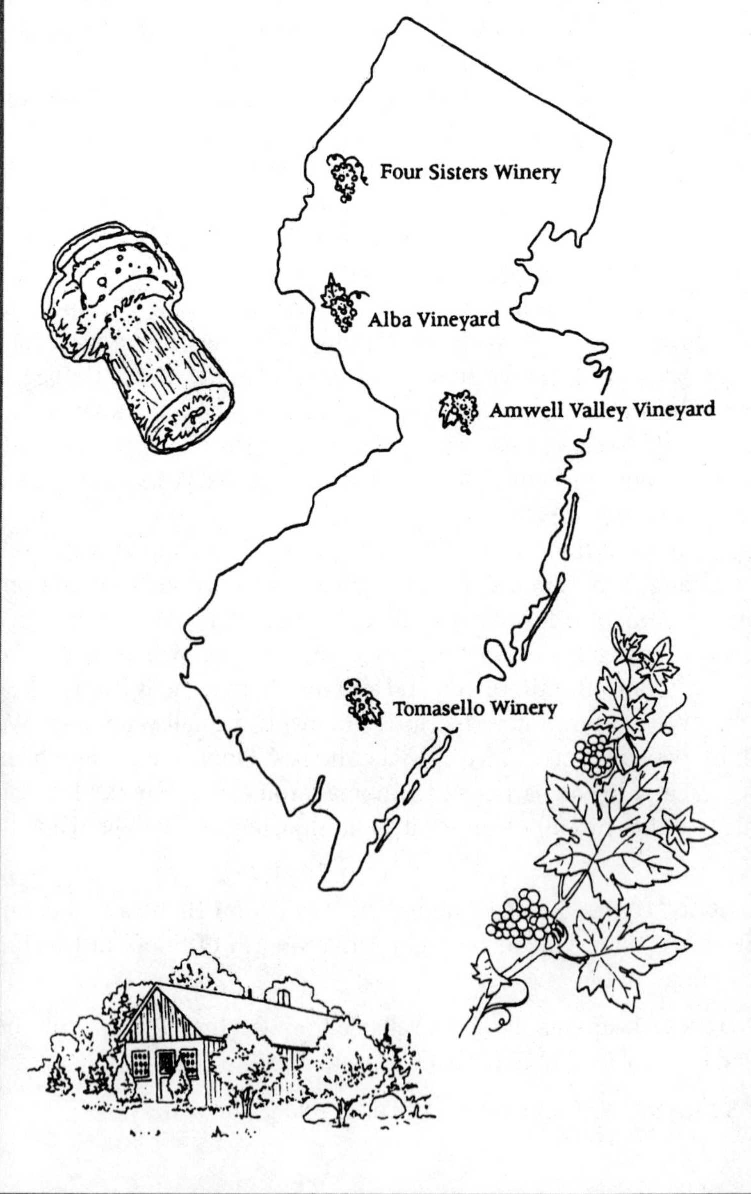
Four Sisters Winery
Alba Vineyard
Amwell Valley Vineyard
Tomasello Winery

Only within the last ten years has the wine industry in New Jersey taken off. Much of the credit goes to the Hunterdon County Grape Growers Association, formed in 1980. Organized to promote wine growing and viticulture, it helped pass the 1981 farm winery bill. Until 1981 the state allowed only one winery per million residents. With the new law this restriction has been changed. In addition, wineries now pay a lower license fee and state tax, making it financially possible for new small wineries to be organized.

New Jersey wineries produce wine from three types of grapes: native American, French-American hybrids, and European vinifera. The wineries are resurrecting a New Jersey tradition begun in the 1700s, when colonists grew wine grapes for the British Empire. Their efforts met with success, and by 1767 two vineyards were recognized by the Royal Society, a group that promoted trade in the New World.

In the mid-1800s, French immigrant Paul Prevost, a grape grower and winemaker, received recognition for his work in cross-pollinating varieties of grapes. Some years later Thomas B. Welch began his fresh grape juice business in Vineland, N.J. By the late 1880s, however, a grape rot attacked the vineyards, and he closed his business and moved to New York.

Prohibition put a stop to most of the winery activity in the state. But now, with the farm winery act, the grape growers in New Jersey hope to surpass the acclaim that their predecessors received more than two hundred years ago.

Alba Vineyard

269 Route 627
Milford, NJ 08848; (908) 995–7800

An old dairy farm approximately 150 years old houses the winery at Alba. Rudolf Marchesi, the owner, worked hard to leave as much as possible of the original barn and its atmosphere. He did a good job. The bottom half of the barn, built into a hill, consists of rough rocky limestone walls. Wood beams and wood ceilings in the upper floor complement the oak aging barrels.

The tour takes you throughout the building, but begins up at the pressing deck. "It's where the stemmer-crusher is," says Jean M. Lynn, office manager and general if-it-needs-to-be-done-I-do-it person at the winery.

In the barrel room are French, Italian, and American oak barrels for aging. Past the barrel room is the main wine cellar, where fermentation and storing occur. You'll see the fermentation tanks and find out why Alba uses fiberglass tanks with floating lids. The winery also uses stainless-steel tanks. Most of these are in the cold room, where the juice can be cold-stabilized to halt fermentation from the wild yeast found on the grapes. Then the tour goes into the bottling room where one machine handles the entire process, including labeling and putting on the capsules, at the rate of a case a minute.

All the wines Alba has available are available for tasting. "Our most popular wine is the red raspberry dessert wine," says Jean. "It sells so fast that people think we only produce it for the Christmas holidays and around Valentine's Day." The wine is gone by spring. "That's the only fruit we don't grow here." Another popular wine is the Proprietor's White Reserve, a dry, light white with a flowery aroma. The New Jersey Blush, a semidry, is similar to a Zinfandel but not as sweet.

Some of Alba's other wines are a barrel-fermented Chardonnay, Proprietor's Red Reserve, the champagnes Brut and Blanc de Noir, New Jersey Vidal, and the award-winning Vintage Port. The wines sell for $4.99 for the Blush to $10 for the Vintage Port.

In the tasting room medals and awards fill the walls, and oak barrels cut in half hold the Alba wine for sale. The wines have won more than ninety awards, including the Governor's Cup, the highest award in New Jersey.

You may also walk into the thirty-five acres of mature vineyards or the rest of the seventy-four-acre farm. "There are sites in the vineyards for a picnic," says Jean. "It's really nice up there." Wine is sold by the glass or the bottle. Or if you're in the area during the last weekend of September, Alba has a Medieval Festival with jousting, fencing, dances, food, and wine tours.

DIRECTIONS: From Interstate 70, take exit 7. Drive west on Route 173 for 1.3 miles. At Route 639 turn left. Drive 2.8 miles to the stop sign. At the intersection continue straight, and the road turns into Route 637. After 2.4 miles winery is on the right.

HOURS: Tours and tastings available Tuesday through Friday from noon to 5 P.M., Saturday from 10 A.M. to 6 P.M., and Sunday from noon to 5 P.M.

EXTRAS: Winery sells wine and wine accessories. Picnic facilities.

Amwell Valley Vineyard

80 Old York Road
Ringoes, NJ 08551; (908) 788–5852

"There was nothing here but a cornfield," says Elsa Fisher, who with her husband, Mike, owns the Amwell Valley Winery. "Most people have one building or something. We had to start from the ground up," Elsa says as she surveys the scenic vineyard slopes, the valley, and the mountain beyond.

"We planted in 1978 and were licensed in 1982." Her son, Jeff, who manages the winery, adds, "We were one of the first granted licenses in 1982 under the farm winery act." One other winery was licensed the same day, the first day the new law went into effect.

Elsa believes they have one of the best reasons for starting a winery: "My husband and I always enjoyed drinking wine. We made some wine at home and liked it."

The winery is open on weekends for tours and tastings, but if you're in the area during the week, you may also have a chance to visit. "If they call us, we'll try to accommodate them," says Elsa. "My husband or my son will take them through if they want to see the winery."

Jeff says that when he gives the tours, he starts outside and shows the visitors the vineyards. Then he takes them inside the winery to see the oak barrels and temperature-controlled fermentation tanks. "They're dairy tanks hooked up to a compressor outside," he says.

Besides old dairy tanks, the Fishers have tried 60-gallon blue plastic barrels and plastic wine tanks up to 1,000 gallons in size. One of their best red wines, the 1982 Maréchal Foch, was made entirely inside plastic tanks. The nineties and convenience have caught up with them, however. The winery now uses only stainless steel for fermentation and an assortment of American and French white oak barrels for aging.

The Fishers believe that their wine is successful largely because it's estate bottled. "All the wine is from vines on the property. Some people buy grapes from others, but we've never done that," says Elsa. Her husband adds with pride, "We're one of the few that make all the wine from our own grapes. Very few can say that." Even their other fruit wines are "estate bottled." "We make plum wine from our plum tree and peach from our peach tree," says Elsa.

Mike has a positive attitude about his vineyard, too. "I just keep planting them. If they die, we replant." Mike grows both vinifera and French-American hybrids. "I got to know Philip Wagner," says Mike of the Maryland wine grower who championed the planting of French-American hybrids. "He said don't even bother with vinifera." But the Fishers didn't listen, and not enough plants have died to make harvesting their ten-acre vineyard easy. "My son always called picking season grape madness," says Elsa.

In the tasting room you can enjoy some of the 2,500 gallons of wine made each year. Multicolored tile covers the floor and the tasting bar. Wood walls are interrupted by stained-glass windows with depictions of grapes.

The Fishers offer a wide selection of wines. They have the reds Maréchal Foch, Cabernet Sauvignon, and Landot Noir; the whites Chardonnay, Riesling, Aurora, Rayon d'Or, Seyval Blanc, Vidal Blanc, and Villard Blanc; and the dessert wines Peach, Plum, Gewürztraminer, and Ravat. They also produce two champagnes, Blanc de Noir and Blanc de Blanc. Wine sells for $5.50 for the Villard to $6.25 for a 1987 Maréchal Foch. The champagnes sell for $10.

Not only was Amwell Valley Vineyard one of the first in getting its license, it also is among the first when it comes to good wine and excellent hospitality.

DIRECTIONS: From Interstate 78, take exit 17. Drive south on Route 31 for 14.5 miles. At the traffic circle, turn right and follow it around to drive east on Route 514 (Old York Road). Drive 1.2 miles and watch carefully for the Fisher mailbox on the right. Turn right for the winery.

HOURS: Tours and tastings offered Saturday and Sunday from 1 P.M. to 5 P.M., or by appointment. Winery closed New Year's Day, Easter, Thanksgiving, and Christmas.

EXTRAS: Winery sells wine and wine accessories.

Four Sisters Winery

R.R. 03, Box 258
Route 519
Belvidere, NJ 07823; (908) 475–3671

"The owners have four daughters, and we have four different fruit wines, each named after a daughter," says Valerie Tishuk, who works at the Four Sisters Winery. "They have a picture of the fruit on the label, and the daughter signs them. They range from sixteen to nine years old." The wines are Strawberry Serena, Robin's Raspberry, Sadie's Apple, and Cherry Melissa.

Then there's Papa's Red. "It was named after the owner's grandfather," says Valerie. "It was one of our first blends. It goes good with spaghetti." Another wine, named after an owner's mother, is the Merrill Blush.

Matty and Laurie Matarazzo own the winery and also the surrounding 392-acre produce farm. The farm has been around for generations, but the winery began in 1984. The Matarazzos saw the winery as an extension of their plantings of fruits and vegetables. Says Valerie of the farm, "[It's] the most diversified produce farm on the East Coast."

The farm supplies the fruit for the fruit wines. "All of our fruit wines are made from fresh fruit," says Valerie. The Matarazzos also have seventeen acres of French-American hybrids. "If it's a nice day we'll take them outside to the vineyards," says Valerie of the twenty-minute tour. "On weekends we have tours on demand, and you can always get a tasting during the week."

But she says most people opt for another alternative. "There's a gourmet and farm market next door. During the summer most people go next door, buy some cheese and crackers, and then come in here and buy a bottle of wine and sit out on the deck." The deck affords a great view of the Matarazzo fruit and produce growing on the rolling hills.

If you visit the winery on a nontour day, you can wander around the tasting room and learn about the wine sold at Four Sisters. Signs hang above each wine for sale and offer information about it, for example, CHANCELLOR—DRY, RUBY-RED, FULL-BODIED VARIETAL. TRULY A PLEASANT DRINK. GREAT WITH MEATS. SERVE AT ROOM TEMPERATURE.

You may also want to arrange a visit to Four Sisters during one of its many festivals. There's a Strawberry Festival in June, a Native American Pow-wow in July, and a Harvest Festival in October. For more information concerning times and dates, call or write the winery.

The winery's twenty-three wines are available for sampling any day of the week. "We always have all our wines open. Most everybody can find something they like," says Valerie. Raspberry and strawberry are big sellers and are not always available. But you can choose from many others, such as the Chambourcin, Chancellor, Villard Blanc, Niagara, Autumn Rouge, or Cedar Hill Rosé. The award-winning wines sell for $4.20 to $6.95. "We've won nine awards in four years. That's very good for a young winery," Valerie says.

Valerie's also proud of the gift baskets the winery offers for sale. "That's what's nice about having the farm market next door. We do gift baskets on demand," says Valerie. You can walk next door, pick out the cheese, jellies, meat, or produce you want included in a gift basket and then go back to the winery and select the wine. "We can put it together and make it. We will do anything you want us to do." With that kind of attitude, you can see why Four Sisters Winery is so popular.

DIRECTIONS: From Interstate 80, take exit 12. Drive south on Route 521. Winery is on the right 6 miles past the intersection of 521 and Route 519.

HOURS: Tastings offered daily from 9 A.M. until 6 P.M. Tours offered Saturday and Sunday year-round from 11 A.M. until 6 P.M. Winery closed New Year's Day, Thanksgiving, and Christmas.

EXTRAS: Winery sells wine and wine accessories. Picnic facilities.

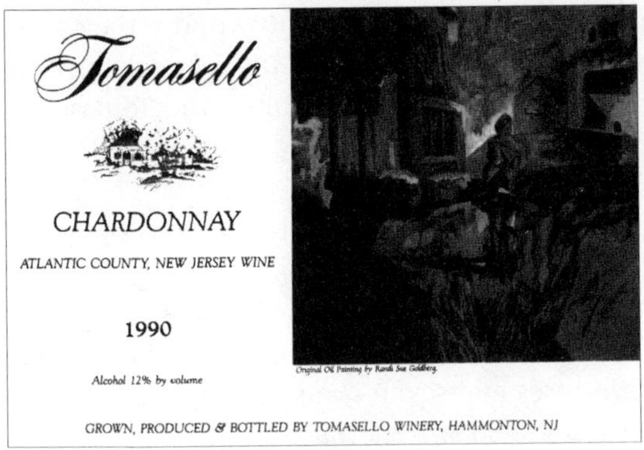

Tomasello Winery

225 White Horse Pike
Hammonton, NJ 08037; (609) 561–0567 or (800) 666–WINE

"The winery was started by my father-in-law. Everyone told him he made the best wine in town. He grew grapes here," says Peggy Tomasello of Frank Tomasello, the founder. "We've been in business since 1933, right after Prohibition. It's third generation, a family operation."

The winery today is on the same location where Frank, a farmer by trade, opened Tomasello Winery in 1933. "Charles and Joseph, his two sons, joined him in business and expanded some," says Peggy, who's married to Charles. During the 1970s the winery expanded operations and added French-American hybrids to the vineyards.

Joseph's wife, Jane, also took part in the winery and helped with the expansion. It was during this time that the groundwork was laid to make Tomasello the largest New Jersey winery, producing almost 100,000 gallons of wine a year. "That's when we started with four or five wines and expanded to more," says Peggy.

Then the third generation came along. "Charles's sons also joined the business, Charles, Jr., and Jack. They're more formally educated," says Peggy. But they weren't pushed to take up winemaking, Peggy is quick to add. "We never encouraged our boys to get into the business. They liked to come over here and get dirty."

With all the wine they make, no wonder they get dirty. "Eight champagnes and twenty still wines. It's very seldom that someone's not satisfied," says Peggy. Of the whites, the Tomasellos offer Cape May, Sauvignon Blanc, Rhine, Niagara, Seyval Blanc, and Ranier White. They also produce a Chardonnay. "They're New Jersey Chardonnay grapes," says Peggy with pride. "They're the only grape that's hand picked. They're real delicate."

On the red side there are the Chambourcin, Cabernet Sauvignon, Claret, Burgundy, and the semisweet Ranier Red. Tomasello produces two rosés and seven champagnes. All the champagnes are made in the traditional *méthode champenoise.*

"We introduced 100-percent fruit raspberry wine," says Peggy. For the holiday season they offer mulled spice wine. Another specialty wine is the American Almonique, a white table wine with almond and other natural flavors. Peggy says, "It's not very syrupy. The ladies tell me they use it in fruit salad. And over pound cake to marinate it." The wines sell for $5 to $15.

Peggy or another member of the family will line up all the wines across the tasting counter from dry to sweet and let you sample as many or as few as you desire. Then, if you've made an appointment or there's someone to cover for her at the sales room, she'll lead you through the large winery.

All along the way she'll explain the still-wine and champagne-making processes. And you can walk behind the winery and see some of the 104 acres of vines used to make the wine.

"There's a lot of facets to the business. You really have to be dedicated and committed," says Peggy. "We're just trying to enlighten the public that New Jersey makes good wine." A regular public relations team all by herself, Peggy will have you convinced after one trip to Tomasello.

DIRECTIONS: From Philadelphia, drive east on the Atlantic City Expressway to exit 28. Drive north on Route 54 for 3 miles. Turn left at Route 30; winery is on the right after .5 mile.

TOURS: Tastings available daily from 10 A.M. until 5 P.M. Tours by appointment. Winery closed Easter and Christmas.

EXTRAS: Winery sells wine and wine accessories.

NEW MEXICO

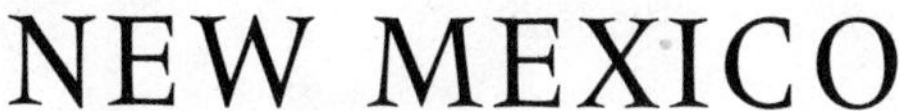

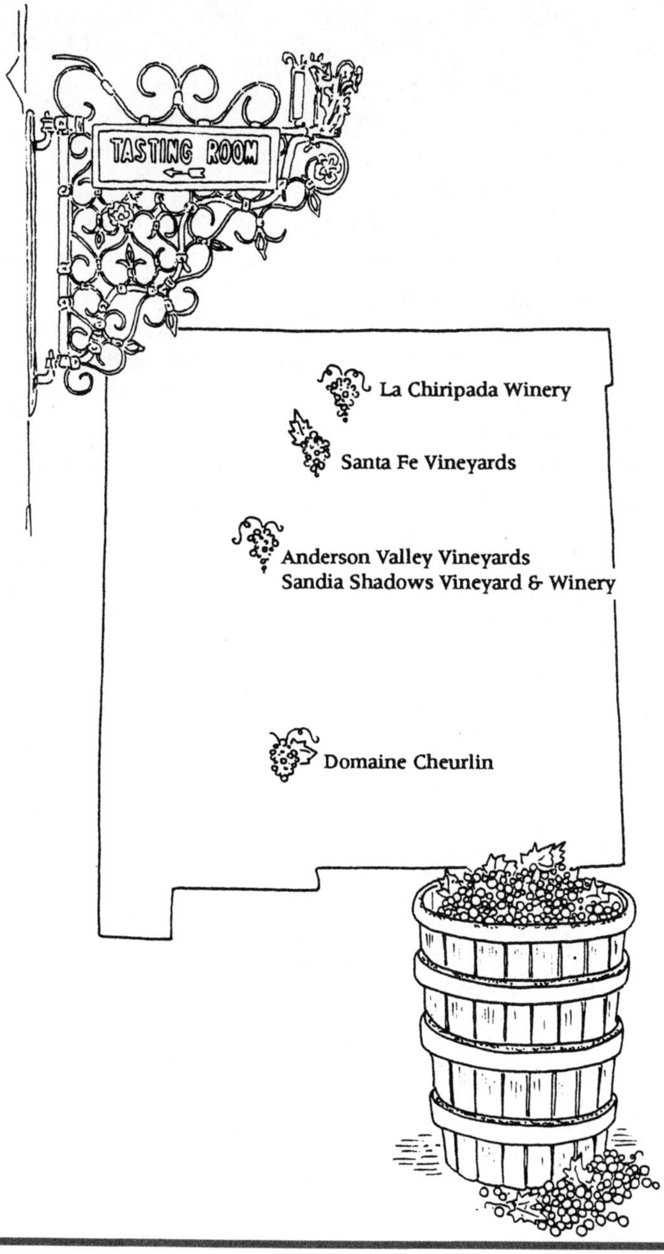

TASTING ROOM
La Chiripada Winery
Santa Fe Vineyards
Anderson Valley Vineyards
Sandia Shadows Vineyard & Winery
Domaine Cheurlin

More than 300 years ago, people were making wine in the area now known as New Mexico. Spanish colonists and mission priests found the climate along the Rio Grande River, which flows south from Colorado through New Mexico, agreeable for grapes.

Some of the first commercial wineries in the United States began in the southern area of the state in the Mesilla Valley. Vineyards continued to grow and spread across the state, and according to the New Mexico Department of Agriculture, "The 1880 census of wine growing reported New Mexico as fifth in the nation in wine production, making 908,500 gallons from 3,150 acres of vineyards." Wineries could be found from Bernalillo in the north all the way south to the Mexican border.

Then nature intervened. The river that had brought good fortune to the vineyards brought destruction. In 1897 floods wiped out many of the vineyards. A drought followed the floods, and by 1914 only eight acres of vineyards were left.

What floods and drought couldn't destroy, Prohibition did. It has taken a long time for the New Mexican wine industry to recover. By 1982 approximately 400 acres had been planted with vines and less than ten wineries were in operation. But less than three years later, the acres of vines increased by more than 1,000 percent. The 1990 harvest produced 700,000 gallons of wine from grapes harvested on more than 4,000 acres.

Vineyards in the north with higher altitudes have been planted using French-American hybrid grapes, while other areas of the state are cultivating the European vinifera vines. The warm days and cool nights keep the grapes growing well. If the grapes continue to thrive, it may be only several years until New Mexico returns to the higher production levels it once knew.

Anderson Valley Vineyards

4920 Rio Grande Boulevard NW
Albuquerque, NM 87107; (505) 344–7266

Festive multi-colored hot-air balloons greet you everywhere at the Anderson Valley Vineyards. In the tasting room you're met by lanterns in the shape of balloons. The winery's label shows a hot-air balloon floating over a vineyard. "We're going to be the official wine of the balloon festival," says Dian Candelaria, the retail manager.

It stands to reason. The owner, Patty Anderson, has had a long association with hot-air balloons. In 1980 Maxie, her husband, and Kristian, her son, made the first nonstop transcontinental balloon flight. The four-day trip covered more than 2,000 miles. Unfortunately, tragedy struck in a later flight. "Maxie died in a ballooning accident over Germany [in 1983]," says Dian.

Patty and Kristian then turned their attention to wine. "They did it as a hobby for years," says Dian. They first planted a vineyard in 1973, but it wasn't until 1984 that Anderson Valley became a licensed commercial winery. Kristian acts as winemaker and winery manager.

The winery's tasting room has a homey atmosphere, with checkered tablecloths, homemade baskets that hang from the ceiling or are filled with wine bottles made into gift packs, and sturdy-looking wood cabinets and armoires—some of them antiques that Patty has collected. It's a pleasant tasting room to browse through while you wait for a tour to start.

"We have tours on the hour when people stop," says Dian. The twenty-minute tour begins at a display of the awards the vineyard has won. "We have four international winners with our wines," says Dian. Then it is back into the part of the building where fermentation takes place. "This is where Maxie used to work on his balloon before we turned it into a winery," says Dian of the large high-ceilinged room. The tour continues into the vineyard.

You can sample the end result while you wait for the tour to start, or you can wait until you've seen the winery. "We give them the wine list and let them pick out the ones they want," says Dian. The wines sell for $6.25 to $15. The winery offers Chardonnay, Chenin Blanc, Johannisberg Riesling, New Mexico Burgundy, Desert Sunset, Muscat Canelli, Rougeon Rosé, White Cabernet, Claret, Sauvignon Blanc, or White Zinfandel.

"The White Zin is so easy, and it's our best seller," says Dian. "The first year we bought some California grapes, but now we're a totally New Mexico product. I think it's gotten real popular to come to New Mexico, and the New Mexican wines have gotten real popular."

DIRECTIONS: In Albuquerque, you can reach the winery from Interstate 40 or Interstate 25. From Interstate 40, west of Interstate 25, exit on Rio Grande Boulevard and drive north; the winery is on the right. From Interstate 25, north of Interstate 40, exit on Montano and drive west to Rio Grande Boulevard. Turn north; the winery is on the right.

HOURS: Tours and tastings available Tuesday through Saturday from noon until 5:30 P.M., and Sunday from 1 P.M. until 4 P.M. Holiday closings vary.

EXTRAS: Gift shop sells wine, wine accessories, and fun unusual items such as wine soap and champagne bubble bath. Picnic facilities.

Domaine Cheurlin

Box 506
Truth or Consequences, NM 87905; (505) 894–3226

There's nothing like having the *méthode champenoise* explained to you by a man with a French accent, especially when you're in such an unlikely place as southern New Mexico, with its arid mesas. But that's what you'll get from Patrice Cheurlin, the owner of Domaine Cheurlin.

His father and brother are still in France making champagne. "We have three wineries in France." So what brought Patrice to the land of adobes and burritos? "In France there's no place to expand. We looked at other places. We were looking in California. As you know everything is so expensive." But cost wasn't everything. "There's no challenge, nothing to prove up there. You know you can make it there."

It wasn't such a sure thing in New Mexico. "We spent two years in analysis. We planted the vineyards in 1981. I start slow. We built the winery in 1985," says Patrice. And the sparkling wines have been doing well. "I expand every year. After two or three years, I am [already] in twenty-five states. Also, Japan and Taiwan. And soon Germany. It's something special because nobody expects it here."

When you tour the 90,000-gallon-capacity winery, you can see where Patrice has added another section onto the main building each year. "I'm small, but I'll be bigger every year. We try to make things better," he says.

But Patrice has had some problems in his 100-acre vineyard. "It got [to be] seven degrees [in 1987]. We lost 70 percent." Usually the problem is heat. To avoid the heat, Patrice has adopted several measures. "We harvest at night, from seven in the evening to 8 A.M. In three days we finish the harvest." Then it's a mad dash to get the grapes to the winery before they lose their flavor. "We leave the grapes in the harvester machine for thirty minutes," explains Patrice. Then they are brought into the winery.

"I have forty-five acres with seven kinds of grapes. This is my experimentation. The soil is very good." Patrice has had success with Pinot Blanc, Pinot Noir, and Chardonnay. With these grapes he makes brut extra dry, brut, and rosé, which retail for $9.50 to $12. He also makes a nonsparkling wine, an Elephant Butte Chardonnay, which sells for $7. "We got a bronze medal in San Diego for the brut. We have won awards from everywhere," says Patrice with obvious pride.

You can taste the award winners after the tour. "Technically I'm supposed to charge [$1 per glass] but I haven't charged yet," says Patrice. Hurry and visit the winery before Patrice loses his accent and before the winery gets so big and impersonal that it collects the fee.

DIRECTIONS: From Interstate 25, take the Truth or Consequences exit. Drive 1.5 miles toward the city. At the only stop light, turn left onto Third, also known as State Highway 51. Drive 15.5 miles east. At the sign, turn left onto the dirt road; winery is on the right after 2.6 miles.

HOURS: Tours and tastings Monday through Friday from 8 A.M. until 4 P.M. Available by appointment on weekends and holidays.

EXTRAS: Winery sells wine and wine accessories.

La Chiripada Winery

Box 191
Dixon, NM 87527; (505) 579–4437

Although Dixon appears to be a sleepy little town, it's also on the main route between the tourist towns of Santa Fe and Taos. "We sell a lot of wine right out of here," says Mike Johnson, co-owner with his brother, Pat. "It's quite a tourist trade."

But the winery doesn't feel like a tourist trap. You drive down a gravel road through rows of vines to get to the winery. Then you go under a white stucco archway with the winery's name and purple grape bunches painted on it. The Southwest-style white stucco winery has plenty of windows on the south side to give it a light, airy feel. A bay window is filled with pottery of blue and brown glazes. The pottery includes wine decanters and glasses, nativity scenes, candlestick holders, and earrings.

"Pat and his wife, Michele, are the potters," says Mike. "This is Pat's place here." Mike lives down the road from the winery.

"We grow some French hybrid varieties. We have an experimental plot," says Mike. The winery first planted vines in 1977 and became bonded in 1981. Mike and Pat are still searching for vines that will grow well in the Dixon climate. The winery doesn't have the best of conditions for grapes. "We have a short growing season here, from May 15 to the 21st of September. We're talking 120 days between killing frosts. Or at best 160 days. We have bud kill almost every year," says Mike. Bud kill refers to when the buds on the vines are killed due to a late frost. "I think we're going to bury the vines [to protect them from the cold]."

But not all is bleak in Dixon. "We just started planting Vidal Blanc," says Mike. "That sure makes a nice wine. Riesling does well. I think the Riesling goes real well with Northern New Mexico cuisine."

The winery has ten acres of vines. It produces 6,000 gallons of wine from its own grapes and grapes from other vineyards. Its award-winning Special Reserve Riesling, made in a German style, comes from its own vineyards. Other wines include the white Vidal Blanc, the blend of Vidal and Villard called Primavera, the blush Vino Sonrojo, Rio Embudo Red, and a dessert port. The wines sell for $4.50 to $9.50.

Although there are no official tours, if you visit the winery when it is not too busy, the staff will be glad to show you around. During the tasting you'll receive expert advice. "I figure out where they're at and talk to them about it. Or I'll suggest," says Mike. "If people want some direction, I'll give it to them. 'What do you drink at home?' And I'll know what to pour them." A nice change from wineries that just ask you what you want to drink.

DIRECTIONS: From Santa Fe take Highway 68 north, or from Taos take 68 south to 75. Head east toward Dixon; winery is on the left after 2.5 miles.

HOURS: Tastings and tours Monday through Saturday from 10 A.M. until 5 P.M. Closed Thanksgiving and Christmas.

EXTRAS: Winery sells wine, wine accessories, and pottery.

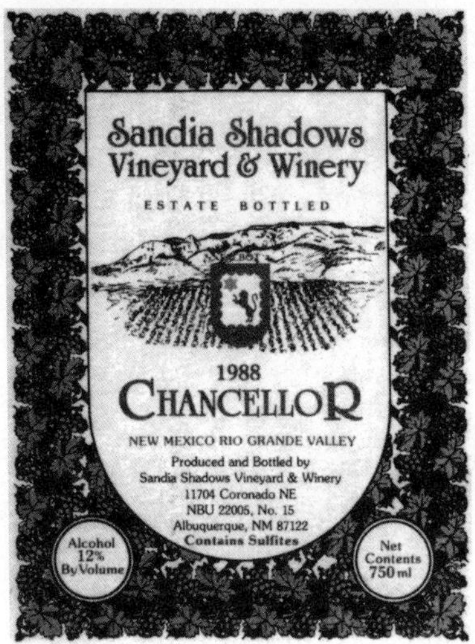

Sandia Shadows Vineyard & Winery

11704 Coronado NE
Albuquerque, NM 87122; (505) 298–8826

"We are a small winery," says Barbara Talbot, who owns the winery with her husband, Lyle. Their winery sits on sixteen acres, fourteen of that in vines, under the shadows of the Sandia Mountains. They first planted a vineyard in 1981 and added the winery in 1984.

Even though the winery is small, Barbara says, "Over the last two or three years, we've grown rapidly." She says it's because more people are trying the wine. "Once they taste it they like it. First thing is to get them to taste it. It's a matter of education." It's a problem that she says is statewide.

On a tour Barbara will show you the lab, where grapes and wine are tested for sugar content, acidity, and other things needed to make good wine. "With the help of a consultant, I am the winemaker, because somebody has to take the information and put it all together," says Barbara.

She puts it all together in estate-bottled wine, which means the grapes are grown right at the winery. "I think that basically we have as good a climate and soil that we could ask for," says Barbara.

Lyle and Barbara make a Chardonnay, which is priced at $8.50; Sandia Blush, at $6.50; Fumé Blanc, at $7.50; Chancellor, at $7.50; and Coronado's Gold, a sweet white blend of Muscat and Seyval Blanc, at $12.

You may drink a bottle while sitting at the picnic table in the grape arbor. Don't miss the street lights located at the end of the vineyard. They are replicas of the lights found in Old Town in Albuquerque. The winery building is a wood A-frame of cream stucco, with an underground cellar to keep the wine at a cool, even temperature.

Things heat up at the winery the second weekend of October, the final weekend of the annual hot-air balloon festival. "I put up a big tent, have an exotic dancer and a jazz quintet," says Barbara. Admission is $3, which includes a wine glass. While you sample the wine at the festival, you may stroll among the food and art booths. The Fall Festival, a musical celebration of harvest, runs Saturday and Sunday from 1 P.M. until 5 P.M.

DIRECTIONS: From Albuquerque, drive east on Interstate 40. Take exit 167, Tramway Boulevard, and turn left. Drive north on Tramway for 5 miles. At the traffic signal for San Rafael, turn left and drive .6 mile. Turn right onto Lowell and make a quick left onto Coronado. Follow the vines to the winery on the left.

HOURS: Tours and tastings Wednesdays, Thursdays, and Fridays from 1 P.M. until 5 P.M. and Saturdays from 10 A.M. until 5 P.M. Closed major holidays.

EXTRAS: Winery sells wine and wine accessories. Picnic and party facilities.

Santa Fe Vineyards

203 Calle Petaca
Santa Fe, NM 87505; (505) 753–8100

Tourists race between Santa Fe and Taos, anxious to see it all, worried that they might miss something. They will if they race on by Santa Fe Vineyards. In an unassuming building on the highway, your senses will be pleasantly assaulted by a virtual museum of the works of Amado Murilio Peña, Jr., one of the Southwest's foremost artists.

Peña posters and prints cover the walls and a Peña can cover you, too, if you buy a Peña T-shirt. Santa Fe Vineyards' wine also has a Peña label. "We were introduced by a mutual friend," says Leonard Rosingana, Santa Fe's owner and winemaker. "We said we were interested in using a past painting. He [Peña] said, 'Let me do something for you.' It was just good timing. He thought it would be fun, too. He uses our wines a lot at openings."

The label's focal point is, of course, a bunch of grapes. Two angular-faced Native Americans, typical of Peña's style, are in the forefront while the background has a pueblo silhouetted against the sun.

Although Len licensed the winery in his first location in 1982, he only moved into this winery in the spring of 1988. "This was an old gas station," says Len. "It wasn't anything until we remodeled." The winery looks nothing like a gas station now, with its earth-tone exterior, its line of windows high on the building's facade, its oak barrels cut in half and filled with flowers decorating the front, and its old steel-wheeled wagon resting by the side of the door. "We got a vintage wagon from the old coal mines near Santa Fe."

Not all has been positive. The vineyard has gone from six acres to one. "My vineyard bit the dust several years ago. So I decided to buy. I'm getting my grapes from the southern part of the state," says Len.

But the ups and downs won't stop Len. Before he opened the winery, he had wine running in his blood. His grandparents came to California from northern Italy, an area where wine is greatly appreciated. In 1975 Len and his brother bought the Ruby Hill Winery in Livermore, California. After years of growing grapes and selling them to other wineries, they decided to produce their own wine under the Stony Ridge label. After a 1980 move to Santa Fe ("my wife was from Santa Fe"), he discovered it wasn't long before the grapes called him back to the business. "Really, finding nothing else I wanted to get into, I went back to the wine business."

His "wine business" includes producing an excellent Santa Fe White Zinfandel and Chardonnay. Or if your tastes run to reds, there's a Cabernet Sauvignon for $12.50. The Viña del Sol blends Seyval and Chenin Blanc for a crisp, fruity wine. The Tinto del Sol blends Ruby Cabernet and Cabernet Franc grapes in a Beaujolais style. The Viña del Sol and Tinto del Sol are available for $5.50. He also makes a Chardonnay for $10.

"I still have the people in mind that I'm doing the wine for. We're producing good quality wine for the money," says Len. "You'd like everyone who walks through the door to take a bottle home." Better yet, buy several bottles to take home and one to enjoy at the vineyard's picnic area. The picnic facilities on the north side of the winery offer a quiet, shaded spot to enjoy a break from sightseeing, or to just enjoy a bottle of good wine.

DIRECTIONS: From Santa Fe, drive north on Highway 285 approximately 16 miles. When you reach the Highway 503/4 (Los

Alamos) turnoff, continue on 285 north for 2.5 miles; winery will be on the right.

HOURS: Open daily during the summer from noon until 5 P.M. for tours and tastings. Winter hours noon until 5 P.M. Friday, Saturday, and Sunday. Winery considering changing hours, so you should call before visiting. Closed New Year's Day, Easter, and Christmas.

EXTRAS: Wine, wine accessories, T-shirts, and Santa Fe spices available at the winery. Picnic facilities.

NEW YORK

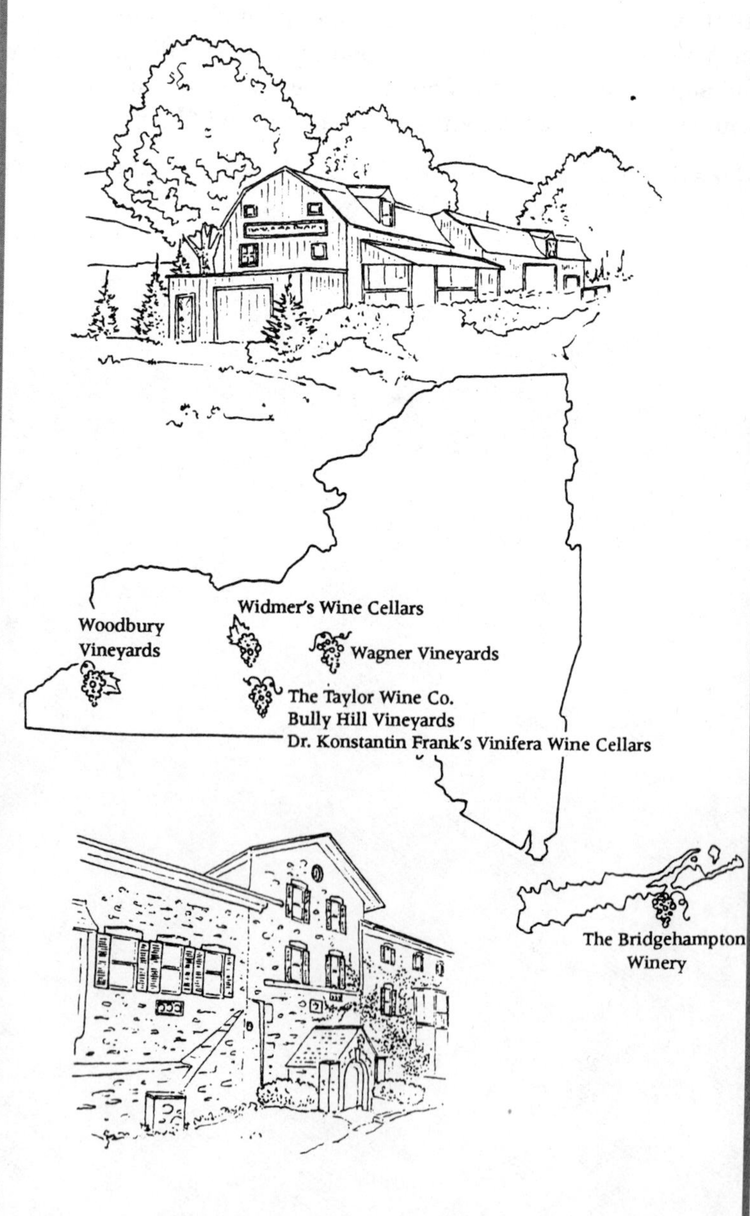

Woodbury
Vineyards
Widmer's Wine Cellars
Wagner Vineyards
The Taylor Wine Co.
Bully Hill Vineyards
Dr. Konstantin Frank's Vinifera Wine Cellars
The Bridgehampton
Winery

Second only to California in U.S. wine production, New York has a long, proud tradition of winemaking. Native grapes thrived in Eastern North America, so much so that Vikings who visited the region called it Vineland.

Some historians have found information that shows French immigrants made wine from native grapes in the Hudson River Valley region as long ago as the 1600s. It was also there that one of the first commercial U.S. wineries opened. The Brotherhood Winery traces its history to 1839. Other wineries soon followed.

While many tried European varieties during this time, New York's claim to fame, and its large production, came from native vines such as Concord and Niagara. Wine grapes are grown throughout the state. The Finger Lakes appellation, or wine growing area, produces more than half of New York's wine. Other federally recognized growing regions are Lake Erie, Hudson River Region, Cayuga Lake, Long Island's North Fork, and The Hamptons on Long Island.

In the last twenty years, New York has increased the number of varieties it produces. You can now find not only native varieties but also French-American hybrids and European vinifera. According to the New York Wine and Grape Foundation, since 1980 vinifera plantings have more than quadrupled. Much of the success of the vinifera comes from the work of Dr. Konstantin Frank. Born in the former Soviet Union, Frank believed vinifera had failed to succeed in the state not because of the climate but because grape growers were not treating the vines properly or using the right variety.

Until recently only a handful of large wineries were responsible for almost all of New York's wine. But in 1976 the state passed the Farm Winery Act, which made it financially possible for small wineries, those that produced less than 50,000 gallons a year, to operate.

Today almost a hundred wineries produce 30 million gallons of wine each year. According to the New York Wine and Grape Foundation, these wineries draw more than half a million tourists annually. New York wineries, with their combination of the old and the new and with their diversity of locations, make New York a great state for winery visitors.

The Bridgehampton Winery

P.O. Box 979
Bridgehampton, NY 11932; (516) 537–3155

"People thought he was crazy," says winemaker Richard Olsen-Harbich of Lyle Greenfield, owner of Bridgehampton. Lyle built one of the first wineries on Long Island in New York. "He fell in love with the idea. Planted them [the vines] all by hand. He just gambled," says Rich. "The gamble has paid off. In the last three years, it's really taken off."

Lyle began planting the vines "by hand" in 1979, and the winery became licensed in 1983. He works in New York City in advertising, but he doesn't let that keep him from the winery during tour time. "The owner gives the tour. With him it usually lasts an hour," says Rich. The tour takes you to the winery's deck, located in the back of the winery, with a great view of the vineyard-covered rolling hills and surrounding land. You may want to visit the winery later in the afternoon. "Watching the sunset out here, it's really beautiful," says Rich. "You really get a sense of history sitting out here and seeing the vines.

"Most of the big fields down here were potato fields. That was the big industry until a few years ago," says Rich. The winery building has also taken something from potatocs. When you drive into the winery parking lot, you see what looks like a hill with a roof plopped on top. Then you walk up steps to enter, and you can see that the building is built into a mound. "That's the style of the old potato farms," says Rich. But the comparison ends there. The rest of the building and the winery is classy, nothing like a potato cellar.

The tasting room has plenty of open space with high ceilings and ceiling fans. The main decorations are posters of the winery's celebrated labels. Designed by Joan Greenfield, the labels feature stylistic paintings by Gretchen Dow Simpson. Her paintings are so elegant they were placed in the permanent collection of the Metropolitan Museum of Art.

The wines have been picking up some classy awards, too. Try the Chardonnay, a consistent award winner, or the renowned Merlot. Other wines that have received high praise are the Riesling, Sauvignon Blanc, and the Premium Cuvée Blanc. Wines sell for $6.99 to $15.99. If the awards are a true reflection of the area's winemaking potential, Long Island has a great future. "California has a history that we're trying to create right now," says Rich. Don't miss being a part of it while it's still young.

DIRECTIONS: From Bridgehampton, turn north at the Bridgehampton Monument and drive for .7 mile. Winery is on the right.

HOURS: Tastings available Monday through Saturday from 11 A.M. until 6 P.M. and Sundays from noon to 5 P.M. Tours available daily on the hour May through Labor Day, and the rest of the year on weekends. Winery closed February, New Year's Day, Easter, Thanksgiving, and Christmas.

EXTRAS: Wine and wine accessories sold.

Bully Hill Vineyards

R.D. #2
Hammondsport, NY 14840; (607) 868–3210 or (607) 868–3610

"Mystery ??? Spot." The billboards used to cover the highways. For hundreds of miles you'd see billboards announcing its approach. "Don't Miss the Mystery ??? Spot, 300 miles." Bully Hill is a lot like the "Mystery ??? Spot." You drive up to the winery on Greyton H. "Blank" Street. Everywhere you look there are signs with names scratched out, boarded over, covered with magic marker, or painted over with black paint. (Even the winery's labels feature the owner's name as Walter S. Blank or Walter S. XXXXXX.) As with the billboards, try as you might, you can't fight the urge to stop and find out what it's all about—all the while fighting the feeling that you're being taken, suckered.

Taking the forty-minute tour, your questions are answered. Walter S. "Blank" is the grandson of the man who founded the Taylor Wine Company in 1878. Walter worked there until 1970 when his criticisms of the way the company was run and the way it made wine became too much for the board of directors. Walter felt the company used too many additives and adulterated the wine with water, sugar, and tank wine shipped from California. After the split he put his full energies into Bully Hill Vineyards, situated on the original site of the Taylor Wine Company, high above Keuka Lake in the Finger Lakes region. He used his name, Taylor, on the label.

Trouble came when the Coca-Cola Company bought Taylor Wine Company and sued Walter over the use of the Taylor name. Coca-Cola won the use of the Taylor name, and Walter was left with a warehouse of wine that he couldn't sell because it had *Taylor* on the label. He instigated a party of all his friends and workers, and armed with magic markers, they attacked the wine labels, scratching out Walter's last name. These labels became so famous, or infamous, that a tradition was born. Walter went crazy scratching out the Taylor name all over his winery, and he painted portraits of his Taylor ancestors with black Lone Ranger masks, hiding their identity. (You can see the portraits on the tour.)

The Lone Ranger comparison rings true. He sees the situation as a battle of good versus evil.

As with the Lone Ranger, good always triumphs. Walter did, too. Instead of hurting him, removing the Taylor name helped. Sales increased, and people flock to the winery to see where the maverick lives. You may get a chance to see him. Walter's known for dropping in and tasting wine, or autographing labels—all designed by himself. The labels feature such eccentric illustrations as a bull dog driving a car or a raccoon in a tuxedo. Wines include Goat White, Sweet Walter, Fisherman's Rosé, and Fish Market White. He produces more than 180,000 gallons of fifty varieties of red, white, and rosé, from dry to sweet. All wines are priced at $5 except for the estate-bottled wine, which sells for $7.

Perhaps unlike the "Mystery ??? Spot," at Bully Hill you won't get taken, except for a good tour. And you may get a look at a man who considers himself a David among a world of Goliaths.

DIRECTIONS: Take Route 54A through Hammondsport. Leaving the village, take the left fork of the Y in the road. Follow the signs to the winery.

HOURS: Tastings and tours available Monday through Saturday from 10 A.M. until 4 P.M. and Sunday from 11 A.M. to 5 P.M. Winery closed New Year's Day, Easter, Thanksgiving, and Christmas.

EXTRAS: Winery sells wine and wine accessories.

Dr. Konstantin Frank's Vinifera Wine Cellars

9749 Middle Road
Hammondsport, NY 14840; (607) 868–4884

If historical markers were put up for places where wine history occurred, Dr. Frank's Vinifera Wine Cellars would probably have the largest marker on the East Coast. It's known as the birthplace of European vinifera in the East.

Dr. Konstantin Frank immigrated to the United States from Europe in 1951. Native grapes and some French-American hybrids were growing in the area, but vinifera had met with failure so often that it was assumed it couldn't grow in the harsh climates of the northern East Coast. Frank didn't go along with that idea; he believed it was a question of finding the right plant, one hardy enough to survive the killing frosts and cold. Frank had experience with winter hardy varieties from his years in the Ukraine.

First for Gold Seal Winery and then for himself, Frank planted vinifera. His first vintage in 1962 was like a stone dropped in a pond—it caused ripples throughout the East Coast vineyards. But while Frank worked miracles in the vineyard, he had some problems with making and selling wine. He preferred experimenting with new varieties and rootstock to blending wine or finding distributors.

His son, Willy Frank, is changing that. He took the reins after his father died several years ago and is concentrating on the wine rather than on creating new rootstock. Willy has his brother-in-law, Walter Volz, helping in the vineyard. Volz had worked and trained with Dr. Frank. There is also an Australian-trained winemaker, Peter Bell.

Vinifera Wine Cellars produces such wines as Johannisberg Riesling, which sells for $7.50, Gewürztraminer, for $12; and Pinot Noir and Cabernet Sauvignon, for $15 and $22 respectively.

Willy's pride and joy, a project he began on his own, is Château Frank, 300 yards down the road from Vinifera Wine Cellars. There he concentrates on making sparkling wines in the traditional French *méthode champenoise.* He grows some of the grapes—Chardonnay, Pinot Noir, Pinot Blanc, and Pinot Meunier—in his own Seneca Lake vineyard. He produces 2,000 to 3,000 cases a year of Château Frank Brut. He is beginning, perhaps, a legacy of his own. Selling for $18, it may be tasted at Dr. Frank's; Château Frank isn't open to visitors.

Dr. Konstantin Frank's Vinifera Wine Cellars offers tours by appointment only. But there are plenty of clippings on the wall from newspapers and magazines to tell you about the Wine Cellars if you didn't make an appointment for a tour. As you sample the wine in the unassuming cement-block building, whoever is pouring will offer plenty of information about Dr. Frank and the wines you're tasting.

If visiting a place where a man made happen what people said couldn't happen is right for you, you'll enjoy Vinifera Wine Cellars.

DIRECTIONS: From Hammondsport, take Route 54A to Route 76. Head north for 6 miles to the winery.

HOURS: Tastings available Monday through Saturday from 9 A.M.

until 5 P.M. and Sunday from noon to 5 P.M. Tours offered by appointment. Winery closed New Year's Day, Easter, Thanksgiving, and Christmas.

EXTRAS: Winery sells wine.

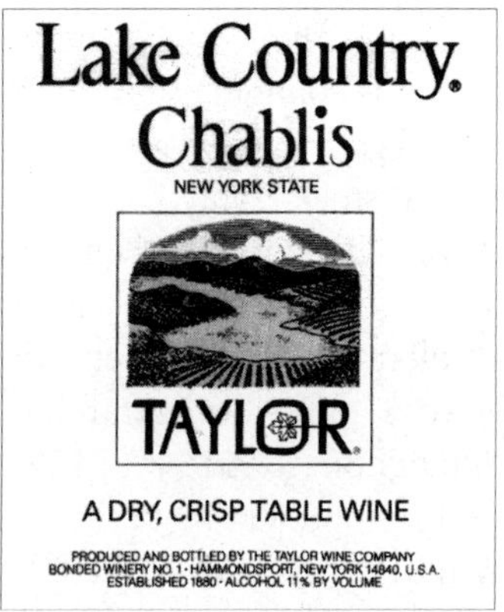

The Taylor Wine Company

Department B
Hammondsport, NY 14840; (607) 569–6292

Perhaps the most impressive aspect about Taylor is its size. Most wineries could fit in one of Taylor's bottling rooms. Thousands of yards, if not miles, of conveyor belts carry bottles up and over other bottles as they are washed, filled, and corked. You'll get to see one of the bottling buildings on your hour-long tour.

The tour takes you by bus to various Taylor buildings such as the champagne building; the grape inspection area, where the grapes are received; the original Taylor sherry room; and a fermentation building. One of the few small things at the winery is the size of the tour groups, which means you may have to wait to take a tour.

Along the way you're given an explanation of the winemaking

process from harvesting to bottling. At the end of the tour, you may taste five wines. The wines sell for $2.99 to $12.99. Then it's back on the bus to the visitors' center, where, after passing through the gift shop, you may taste two more of Taylor's wines. If you prefer, you may skip the tour and still receive a tasting of five wines at the 66-foot, horseshoe-shaped tasting bar in the visitors' center.

Taylor also offers a twelve-minute film that is shown every twenty minutes. It's a great introduction to winemaking in the Finger Lakes area and a good way to spend twelve minutes while you're waiting for your tour to begin. The "theater in a wine tank" is a forty-seat auditorium built in a 35,000-gallon redwood cask that has been filled with chairs instead of wine. The film also covers Taylor's history. Begun in 1880 by Walter Taylor, it has grown to be one of the largest in the world. The Taylor Wine Company also produces Great Western, Gold Seal, and Lake Country wines.

A walk around the interior of the visitors' center is another hour-tour itself. Antique equipment used in the winemaking process and old photographs of the area's history are fascinating. One section of the room holds original copies of newspapers proclaiming the arrival of Prohibition and the coming of repeal. The *Hammondsport Herald* Special Repeal Edition, dated December 7, 1933, features stories of all the local wineries and how prepared they were to resume operations.

Taylor Wine Company remained open during Prohibition by selling sacramental wines and grape juice. Along with a few other wineries in the United States, Taylor sold grape juice with specific instructions on how *not* to turn the grape juice into wine.

If you're looking for a winery that does everything in a big way, don't miss Taylor's.

DIRECTIONS: On Route 17, take exit 38 at Bath. Drive north on Route 54 approximately 5 miles. Turn left onto Pleasant Valley. After .5 mile winery is on the right.

HOURS: Tasting and tours available January through March, Monday through Saturday from 10 A.M. until 4 P.M. Open daily April through December from 10 A.M. until 4 P.M.

EXTRAS: Winery sells wine, wine accessories, and gifts.

Wagner Vineyards

9322 Route 414
Lodi, NY 14860; (607) 582–6450

Picture-perfect describes the winery at Wagner Vineyards. The octagon-shaped building set atop a hill, overlooks the vineyards rolling toward Seneca Lake and the hills beyond. Bill Wagner, the owner of the winery that opened in 1979, built the structure with a vast expanse of windows to take advantage of the view.

You'll see the whole building on a tour, which begins outside where the grapes are received and the pressing takes place. Then you move inside to the tank area, the bottling line, and the wine cellar, which is filled with oak barrels and casks. The winery's distinctive design has a solid earth core (in the middle, like a donut), which keeps the wine cellar naturally cool.

At the tasting room you'll be invited to sample a large selection of wines. The wines open for tasting change monthly, a perfect reason to visit Wagner often. During the tasting the guide explains how to taste wine and gives information on each wine being tasted. After the complimentary tasting you may sample more wines for a small fee.

Wagner offers a wide variety of choices, ranging from drier

wines such as Chardonnay, Seyval Blanc, Gewürztraminer, Johannisberg Riesling, DeChaunac, and Pinot Noir, to semisweet wines such as Niagara, Delaware, and any of the Alta wines. The Alta wines—Alta B, Alta Blanc, and Alta Blush—were named after Bill's mother, Alta B. Wagner. The winery also produces three types of grape juice: Blush, Ravat, and Johannisberg Riesling.

In addition to the beautiful setting and architecturally interesting winery, Wagner has a restaurant. "People today have such a strong interest in wine and food," says Bill. "The combinations are endless, and we just had to have a place to showcase our wine and food ideas. Hence, the Ginny Lee Cafe was created." The cafe is the namesake of his granddaughter, Virginia "Ginny" Lee. The glass-and-wood dining area is spacious and offers a spectacular view of Seneca Lake. You can get an even closer view by enjoying your food and wine outside on the deck.

The cafe offers lunch and Sunday Brunch with prices ranging from $4 to $10; dinners cost from $7 to $15 per entree. Call the cafe for daily specials or current dining hours (607–582– 6574).

DIRECTIONS: From Watkins Glen, head north 14 miles on Route 414 to the winery; from Lodi head south on Route 414 for 3.5 miles.

HOURS: Tours and tastings available Monday through Friday from 10 A.M. until 4:30 P.M. and Saturday and Sunday from 10 A.M. to 5 P.M. Winery closed New Year's Day, Thanksgiving, and Christmas.

EXTRAS: Winery sells wine and wine accessories. Picnic facilities.

Widmer's Wine Cellars

One Lake Niagara Lane
Naples, NY 14512; (716) 374–6311

Widmer's Wine Cellars is not only the third largest winery in the United States, producing a million gallons of wine a year, but it's also the third oldest in New York. Widmer's began quietly enough in 1888 with the first vintage. John Jacob Widmer and his wife, Lisette, settled in the hills that reminded him of his Swiss homeland.

Wine was such a part of his life that he built his first house with a wine cellar. John Jacob set to planting vines and making wine for his family and neighbors. As more people tried his wine, his sales increased. By the late 1880s he had expanded the business and added a stone wine cellar.

The second generation of Widmers sold the company, and today it's owned by the Canandaigua Wine Company. Its large complex of buildings may be seen on the thirty-minute tour. The guides announce that there are 140 steps on the tour, so that those who don't enjoy climbing may just sit back and taste. You end up in the wine shop in John Jacob's wine cellar.

The twenty-minute wine appreciation and tasting shows you how to taste wine to receive its full impact and gives you a nice variety of Widmer wines to sample. Widmer's, with thirty varieties, has six basic groups of wine. The Lake Niagara wines, which count for 60 percent of its business, are from the Niagara grape and made in white, red, and pink and Lake Light and Lake Roselle. Naples Valley wines are blended, such as Chablis, Burgundy, and Blush. There's also a line of champagnes, dessert wines from French hybrids, and the varietals Riesling, Cabernet Sauvignon, and Chardonnay. Most wines sell for $3.99 to $12.99.

Across the street from the wine cellar is the gift shop, which is loaded with wine accessories and glassware. For those who usually shun gift shops, don't miss this one. It sells the most unusual and flavorful wine jellies. The sample tray, set by the door, offers several to try such as Johannisberg Riesling, Port, Burgundy, and Niagara. You'll be tempted to sit there and make it your lunch. For $3.50 a jar, it's a bargain.

DIRECTIONS: From New York City, take Interstate 90 to exit 44. Drive south on Route 21 to Naples. Coming from Route 17, drive north on Interstate 390 to 371 north to 21 north to Naples. Winery is at the north edge of town.

HOURS: Tours and tastings from June through October, Monday through Saturday from 10 A.M. to 4 P.M. and Sunday from 11:30 A.M. to 4:30 P.M. November through May the winery is open daily from 1 P.M. to 4 P.M. Winery closed New Year's Day, Easter, Thanksgiving, and Christmas.

EXTRAS: Gift shop sells wine, grape juice, wine accessories, and gift baskets.

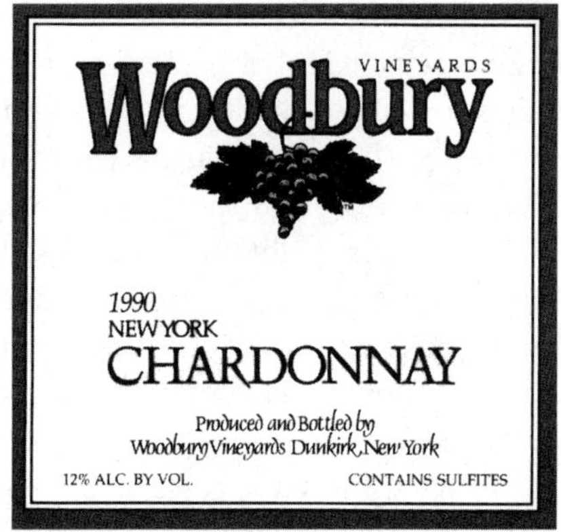

Woodbury Vineyards

South Roberts Road
Dunkirk, NY 14048; (716) 679-9463

Over a barrel about what to give someone you *think* has everything? Give them a barrel. Woodbury Vineyards offers a feature called Adopt a Barrel. It is a way for Woodbury to generate cash now instead of waiting a few years for the wine to mature. "Our Adopt-a-Barrel program is kind of unique," says wine shop manager Virginia "Ginny" Bragg. "For $500 you get your nameplate [on the barrel], visitation rights, and adoption papers."

Barrels hold Chardonnay or Cabernet Sauvignon. "Each year you receive two cases of barrel-aged wine. After four years you can take your barrel with you. In all, you get $800 worth of wine plus the barrel."

When you visit your barrel, you're allowed to taste from it. This gives you a chance to see how the wine tastes at different stages of its aging. "It's been very popular," says Ginny. "People from all over, in New York, out of New York, have adopted barrels."

The Woodbury family began the adoption program as a way to raise capital for the winery that opened in 1980. Now it's so popular, Woodbury fans wouldn't let them discontinue it.

The winery has a hundred-acre vineyard from which it produces approximately 25,000 gallons a year. Most of the vines are in Chardonnay and Riesling. Woodbury produces vinifera wines that have won major awards. Chardonnay, Riesling, Cabernet Sauvignon, blends, *méthode champenoise* sparkling wines, Blanc de Blanc Brut, and Riesling are all priced from $4.99 to $11.99. You may taste the wines after the thirty-minute tour, or you can just drop by for a taste. Harvest is from mid-September to mid-October, and Woodbury's is one of the few wineries in the United States that's happy to have you come and watch the whole crushing process.

Bring your wine questions, Ginny promises you won't be disappointed. "It's very much a family operation," says Ginny. "At least one of the Woodburys is usually here, or our winemaster, Markus Riedlin. If you have a question we can find an answer." A good question would be how Woodbury managed to produce a Chardonnay that won a gold at the 1988 Intervin International Competition at Toronto. Against more than 1,400 wines from France, California, and the rest of the world, only one wine, from Australia, scored higher than Woodbury's Chardonnay. There are plenty of California winemakers that would love to know how to do that.

DIRECTIONS: From the New York State Thruway, take exit 59. Turn south on Route 60 for .7 mile. Turn left, east, on Route 20 for 1.3 miles. Then turn right on South Roberts for .9 mile. At the fork in the road, go left; winery is located in the fork.

HOURS: Tours and tastings available Monday through Saturday from 10 A.M. until 5 P.M. and Sunday from noon to 5 P.M. Closed New Year's Day, Easter, Thanksgiving, and Christmas.

EXTRAS: Winery sells wine and wine accessories. Picnic facilities.

NORTH CAROLINA

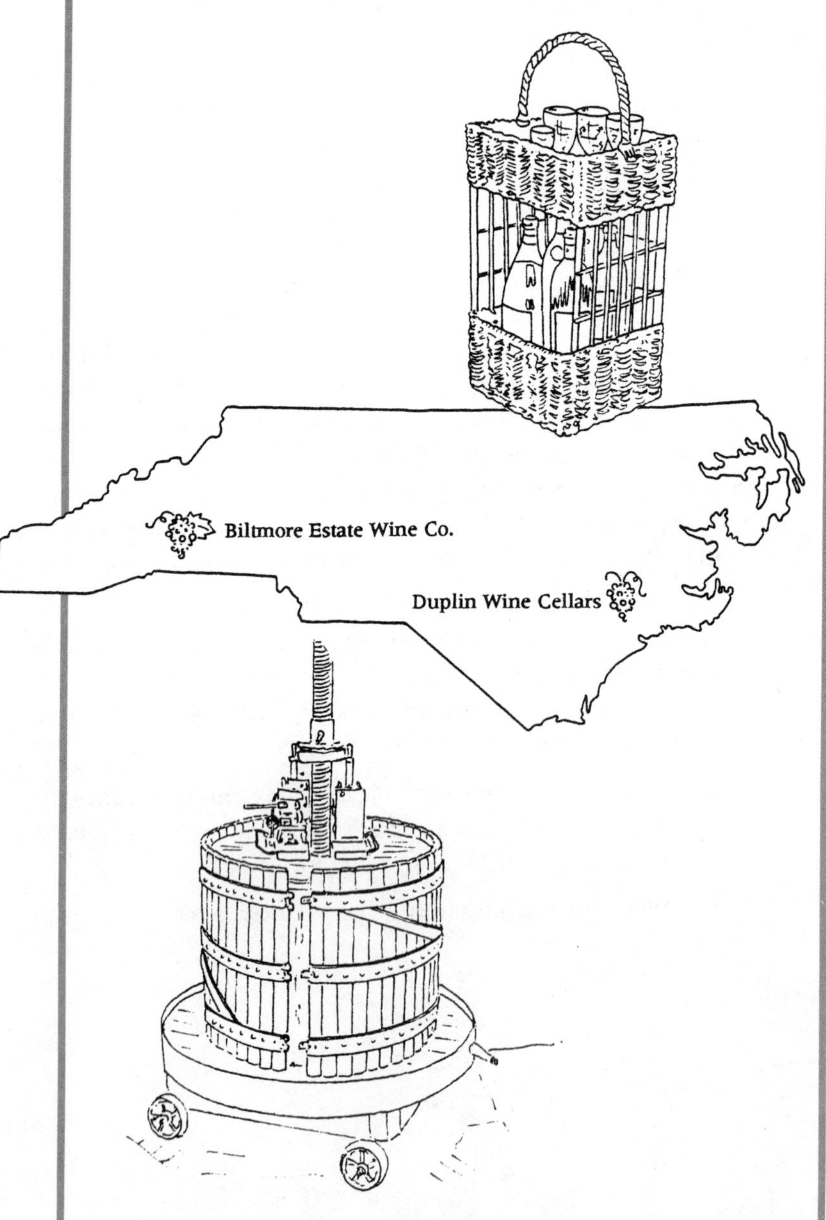

North Carolina has a 200-year-old tradition of grape growing. "God planted Muscadines in North Carolina," says a film shown at Duplin Wine Cellars in Rose Hill, North Carolina. In 1584, when Sir Walter Raleigh explored the North Carolina coast, he found such an abundance of Scuppernongs (a type of Muscadine) that he wrote, "so full of grapes as the very beating and surge of the sea overflowed them. . . . In all the world, the like abundance is not to be found."

Muscadines have been used for winemaking since the 1700s. German-Swiss immigrants produced wine from the grapes, adjusting old European methods to suit the grapes they found in their new land.

In the early 1800s, Sidney Weller, a pioneer in winemaking in North Carolina, sold his wine throughout the area from his winery, the Medoc Vineyard. He became known for producing a sparkling wine from Scuppernongs. By the mid-1800s Medoc Vineyard was joined by several other wineries.

Weller was bought out by the father and uncle of Paul Garrett, a man who in less than forty years owned five wineries in North Carolina and was selling his wine, the famous Virginia Dare, all across the country. During the Prohibition years, Garrett moved his business to the Finger Lakes region in New York.

Up until Prohibition, the state had more than thirty-three wineries. After repeal, a few wineries came and went, such as one in Greensboro and another in Pinehurst. But it took the state until 1976 to recuperate from Prohibition. Duplin Wine Cellars first planted vines in 1972 and had its first vintage in 1976. Less than ten years later, Duplin was winning medals with its wine, proving that North Carolina could indeed produce quality wines.

In 1985 Duplin was joined in winemaking by the Biltmore Estate Wine Company. Biltmore wines have also received numerous awards. With more wineries following in their paths, respect for North Carolina wines will surely grow.

Biltmore Estate Wine Company

One North Pack Square
Asheville, NC 28801; (704) 274–5903

A mansion, a cafe, a restaurant, three gift shops, a carriage house, gardens and greenhouses, ponds and pools, *and* a winery await you at the Biltmore Estate. George Vanderbilt, grandson of the industrialist and financier Cornelius Vanderbilt, had architect Richard Morris Hunt and landscape architect Frederick Law Olmsted plan an American version of a European estate.

The house, originally situated on 125,000 acres, took five years and a thousand workers to build. Vanderbilt named the house and the estate by taking the name of the family's original Dutch village, Bildt, and the Old English word *more,* which meant "rolling hills," and combining the two.

A day-long visit to the estate will allow you time to tour the house, gardens, and winery. The winery is a recent addition. Vineyards were first planted in 1971. In the Biltmore tradition, the vineyard began in a big way with approximately one hundred acres of vines. The winery itself opened in 1985. It produces more than ten types of wine such as Chardonnay, Riesling, Cabernet Sauvignon,

Chenin Blanc, and sparkling wines. The wine sells for $8 for the Biltmore house wine to $16.99 for the sparklings. You have the opportunity to taste several of the wines at the end of the tour.

Depending on your mood, you may have a nine-minute tour or an hour-long one. The tour begins with a nine-minute slide show on the history of wine. After the show you can zip off and taste some wine, or you may walk leisurely through the exhibits and displays on the self-guided tour. Without a doubt, Biltmore offers one of the most elaborate self-guided tours in the United States. There are buttons that light up displays, continuously run videos on champagne making, and overlooks that let you peer down at the stainless-steel fermentation tanks or oak barrels.

The best part of the tour consists of the photographs of the old Biltmore Dairy, which are displayed on the staircase walls as you walk into the cellar. The winery is housed in the old dairy barns, and the tasting room in the old milking room. The dairy began in 1879, and you can see photographs of milkmen, dairy cows, and old milk trucks.

In the middle of the dairy complex is a nineteenth-century European-style clock tower. Originally the clock tower had only three faces; the side pointing toward the cow pastures was without a face, since the cows didn't need to know the time. Years later a fourth clock was added for aesthetic purposes.

Although the winery tour lasts an hour, you'll probably want to plan on spending the day. Admission to the estate (you can't visit just the winery for a reduced fee) costs $21.95 for adults and $16.50 for students and ten- to fifteen-year-olds; children under nine are free. The price includes admission to not only the winery but also the house and the gardens. (Don't miss one of the first heated indoor pools, a bowling alley, and exercise room—all in the basement in the house.)

If you have a day to spend exploring how the wealthy lived at the turn of the century and also to taste some wine, you won't be disappointed with Biltmore Estate Wine Company.

DIRECTIONS: In Asheville from Interstate 40, take exit 50 from the west or exit 50B from the east. Drive north three blocks on Highway 25 and follow the signs to Biltmore Estate.

HOURS: Tours and tastings Monday through Saturday from 11 A.M. until 7 P.M. and Sunday from 1 P.M. until 7 P.M. Last tours begin at 7 P.M., except during January, February, and March, when the winery closes at 6 P.M. Winery closed New Year's Day, Thanksgiving, and Christmas.

EXTRAS: Winery sells wine and wine accessories. Picnic facilities.

Duplin Wine Cellars

P.O. Box 756
Highway 117 North
Rose Hill, NC 28458; (919) 289–3888

"On Saturdays I do the tours and tastes," says Ann Knowles, champagne and wine-jelly maker and manager of the bottling room. "I just have a good time. I like people. I don't meet too many mean people." They wouldn't dare be mean to Ann, with her quiet, soft Carolina drawl and face ready to smile. "We've had people from everywhere—China, Japan, two little old ladies from Yugoslavia. I love it."

Her enthusiasm translates into a fun time at Duplin. You are greeted at the door with a hello and a smile and asked if you want to get right down to tasting or to take a tour first. If you opt for the hour-long tour, you're ushered to the back of the tasting room for a ten-minute video on the history of winemaking in North Carolina and how it's done today. Then Ann whisks you into the back rooms where the Duplin sparkling wines are made. "We moved all our production down the road," says Ann. Now only the champagnes are produced at the tasting room location.

Ann offers one of the most complete tours on the making of champagnes, *méthode champenoise*. But then she should know since the champagne is under her watchful eye. "It takes a year to make wine but two years to make champagne," says Ann. She not only tells you about it but shows you. Ann has kept bottles at various stages of the process so you can see what is accomplished at each one. For example, Ann has a cloudy bottle and a clear one to show you the difference in the wine before and after riddling. Riddling is the process of clearing the wine of sediment and yeast.

Ann says most people's favorite part of the tour is not in the back in the sparkling-wine room but out in front of the winery where there are a few vines. "In the spring we show people how to tell by the leaves what color the grapes are going to grow here. We show them how to tell how old the vines are. People love that," says Ann.

"We started with three wines and now we have about twenty," says Ann as she pours the samples. Some of Duplin's wines are Chablis, Carlos, Magnolia, Scuppernong, and Burgundy. "All of our wines are made from Muscadine grapes." The winery also sells fortified wines. "Our ports and sherries are 20 percent [alcohol]. They have brandy added in them."

Most of the wines sell for $3.89 and the champagne for $7.95. Duplin also sells grape jellies and sparkling nonalcoholic juice, which is one of Ann's favorites. She asks, "Now don't that taste like you just bit down on a grape?"

DIRECTIONS: From Fayetteville, take Highway 24 east for 42.3 miles. Turn right on Highway 117 and then drive south for 12.4 miles. Winery is on the right.

HOURS: Tours and tastings available Monday through Saturday from 9 A.M. until 5 P.M. Winery closed Sunday, New Year's Day, Easter, July 4, Labor Day, Thanksgiving, and Christmas.

EXTRAS: Winery sells wine and wine accessories.

OHIO

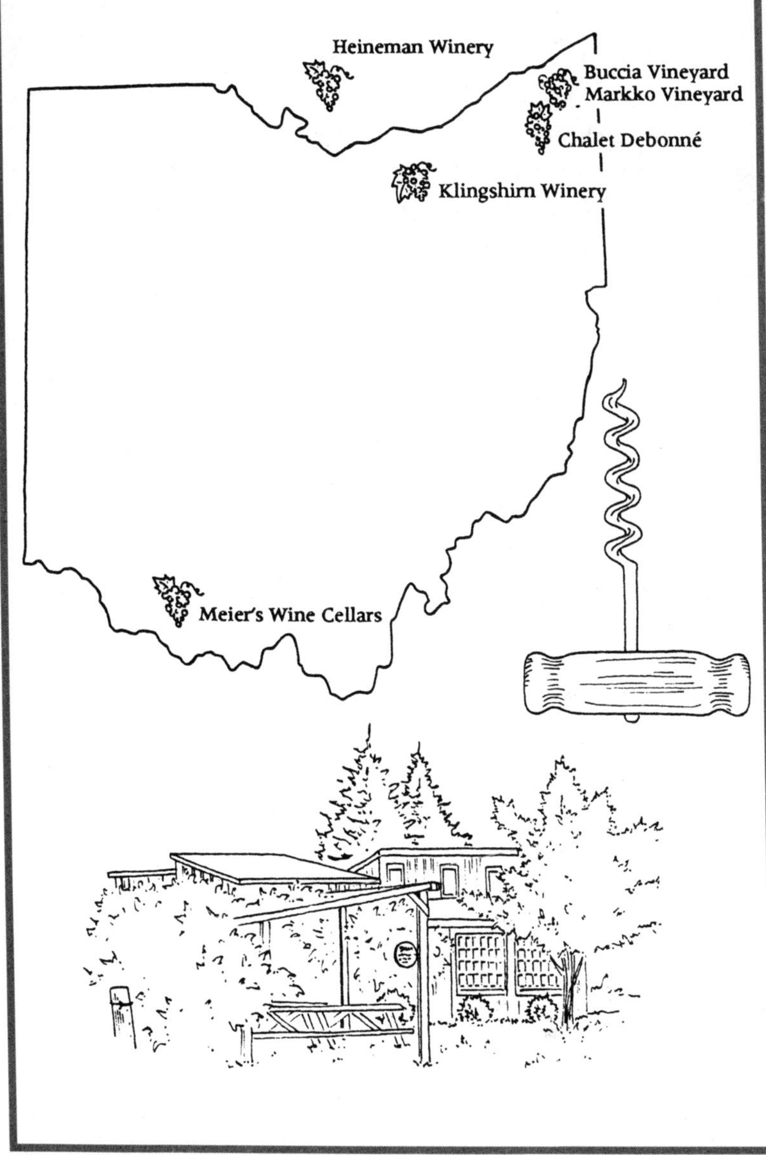
Heineman Winery
Buccia Vineyard
Markko Vineyard
Chalet Debonné
Klingshirn Winery
Meier's Wine Cellars

In the 1820s, Nicholas Longworth received a few grape cuttings from a friend. A lawyer by trade, Nicholas planted the vines and within the next forty years helped develop winemaking in the Cincinnati area. He experimented with many varieties but had most of his success with the Catawba. Longworth's work and that of others soon had people calling Ohio the Rhine of America.

Through the 1840s and 1850s, land devoted to grapes almost doubled each year. By 1850 more than 900 acres were under cultivation. But with the next twenty years came devastation. Land in the area became valuable as the city expanded, more vines were being attacked by diseases, and during the Civil War, with no one to tend them, many vineyards fell to ruin.

While southern Ohio faced decline, the Lake Erie area in northern Ohio took off. The Bass Islands, off the coast of Sandusky, became wine islands. The first island was Kelley's Island. Datus Kelley purchased the island in the 1830s and planted grapes in the 1840s. Within ten years the island had a vineyard of more than a thousand acres. Next came North Bass Island; Put-in-Bay, also called South Bass; and then Middle Bass. By the 1870s the winery on Middle Bass had become the largest winery in the United States, and the state was producing more wine than any other.

Just as the area was booming, Prohibition came along. Almost all the wineries shut down. Some stayed open selling grape juice, but as with many states, Prohibition changed the face of Ohio's wine industry forever. Although not reaching the pre-Prohibition grandeur, it has made a comeback. Many of the diseases that ruined vineyards at the end of the nineteenth century can now be prevented or treated with chemicals. And new winemakers are trying new types of vines suited to the area and using new techniques, continuing Ohio's two-hundred-year-old tradition of winemaking.

Buccia Vineyard

518 Gore Road
Conneaut, OH 44030;(216) 593–5976

"We may very well be the smallest," says Joanna Buccia, who owns Buccia Vineyard with her husband, Fred. That may be why she adds, "We're real relaxed." They're staying away from the pressures and size of a big winery. "We have a storage capacity of 4,000 gallons."

Like many winemakers, Fred started at home. "He always had something brewing in the basement," says Joanna. Then one day he came home from work, and according to Joanna, said, "'Let's move out into the country and grow grapes.' He was a frustrated home winemaker that got carried away. The first year we planted a thousand vines."

In their vineyard they grow Baco Noir, Seyval, and Vignoles, which are French-American hybrids, and they make wine from native grapes. Prices range from $4 to $7 a bottle.

At Buccia Vineyard you can sit back and enjoy your wine or go jump in a hot tub. "It has taken off," says Joanna. "Basically it's like a bed-and-breakfast. We have two rooms." In back of the tasting room are the hot tubs. "You have total privacy. It's more of a get-away than a Holiday Inn." There's also a brick patio outside for cookouts or lounging.

If hot tubs aren't your idea of entertainment, how about singers? "On Saturday night we have a folk singer. It starts at nine. We have fresh bread, salami," Joanna says. And of course, wine. Also, says Joanna, "Weekends are a good time to see grapes being picked and pressed." She or Fred will spend as much time as you want on a tour. "We can spend five minutes or a half hour. If you want to, we'll give you a tour of the vineyards or the wine cellar."

You don't have to worry about bringing your children to this winery. "Bring your kids. We have a playhouse; we have a gym set," says Joanna. There's something for everyone to do. "We have beautiful beaches." She says the fishing is great, too. "We're the Walleye capital." And Joanna will be more than happy to show you how to get to the covered bridges that dot the area—a great place for a picnic if the hot tub is busy.

DIRECTIONS: From Cleveland, take Interstate 90 east to exit 241. Go north on Highway 7 for 2.1 miles. Turn left on Highway 20 for .9 mile and then turn right on Amboy for .7 mile. Turn left onto Gore Road for .3 mile and winery is on the right.

HOURS: Tours and tastings all year Monday through Saturday, but appointments should be made Monday through Friday during the day. Winery closed Sundays, New Year's Day, Easter, Thanksgiving, and Christmas.

EXTRAS: Winery sells wine, wine accessories, and crafts. Picnic facilities.

Chalet Debonné

7743 Doty Road
Madison, OH 44057; (216) 466–3485

Like many Ohio wineries, Chalet Debonné offers a place where you can do more than tour the winery and stand at the tasting bar. If you're looking for late-night wine tasting, you can find it at the Chalet. Wednesday and Friday evening the winery is open till eleven. Appropriate dress is required after eight.

The chalet-like tasting room is designed more like a restaurant. The first room you enter has a large stone fireplace, dark-wood-paneled walls, and tables covered with red-checked tablecloths. A second room off to your left offers a completely different atmosphere, with large floor-to-ceiling windows, ceiling fans, and hanging ferns.

In either room you may sit and enjoy wine and food. The winery offers wine by the glass, ten samples of wine, or a sampler of its premium wines for $3. To complement the wine you may have a cheese, sausage, and homemade bread plate or a deli meat-and-cheese plate. If you'd prefer just a few samples, you may taste those free at the tasting bar, but that isn't what this winery focuses on.

The winery produces more than twenty wines from native

American, French-American, and vinifera grapes. Under the Chalet Debonné label it sells Niagara for $4.15, Delaware for $4.25, Erie Shore Chablis for $4.25, River Rouge for $4.15, and other types of wine. The Debonné Vineyards label goes on its premium varietal wines such as Chardonnay, priced at $8.46; Johannisberg Riesling, priced at $7.46; and the Cabernet Sauvignon, priced at $8.96. Red or white grape juice is available for $2.50.

The fifteen-minute tour will show you where the wines are made. Particularly interesting is the cellar room, where you can see oak barrels made from trees grown on the family farm. Although the winery was licensed only in 1971, the family has been making wine for four generations. Anton Debevc first planted vines, and he was followed by his son, Tony, and then Tony's son, Anthony. Anthony is responsible for much of the chalet's growth and new vineyard plantings of vinifera and French-American varietals.

The Debevcs have created a nice restaurantlike space at the winery. As long as you don't expect to talk with the winemaker or stand at the tasting bar and discuss the latest vintage, you won't be disappointed.

DIRECTIONS: From Cleveland, drive east on Interstate 90 to the Highway 528 exit. Drive south 1.7 miles. Turn left onto Griswold for .8 mile. Turn left on Emerson, which becomes Doty. Drive 1.5 miles; winery is on the right.

HOURS: Tastings available Tuesday through Saturday from noon until 8 P.M. and Wednesday and Friday from 1 P.M. until 11 P.M. During January tastings available from noon to 5 P.M., Tuesday through Saturday. Tours available every half hour beginning at 1:30 P.M. Winery closed Easter, Thanksgiving, and Christmas.

EXTRAS: Winery sells wine, wine accessories, and food.

Heineman Winery

Box 300
Put-in-Bay, OH 43456; (419) 285–2811

Not too many people visit the Heineman Winery during the winter. To get there you'd have to fly. The double-deck auto and passenger ferries stop running to the island in Lake Erie in the fall. But during the summer thousands take the thirty-minute ferry ride to the island resort. Gustav Heineman's trip was a bit longer.

"He came from grape country, so he went to grape country," says Louis Heineman of his grandfather, Gustav, the founder of the winery. "My grandfather came from Germany, Baden province. He went to work at what now is the Lonz Winery. He then went back to Germany. In 1884 he came back to this island. In 1888 he started the winery and ran it until Prohibition. He couldn't see making grape juice."

But Louis' father, Norman, had other ideas. "My dad made grape juice and put it in barrels and shipped it around the Midwest." During Prohibition families were allowed to make wine for personal consumption. Some was made from Heineman's grape juice. "After Prohibition my father went to Toledo and got a grower's permit. My father ran the winery up until the sixties or seventies." Louis made the wine until 1981 when his son took over the job.

You'll probably get the chance to learn the Heineman history firsthand from a Heineman. "In a normal day we'll have at least eighteen tours. I take about five or six groups through a day," says Louis. Don't let his straight-faced delivery stop you from asking questions. Once he sees you're genuinely interested in the winery, Louis will share his wealth of information.

The winery tour lasts approximately twenty minutes and includes a walk through the winery and a description of the wine-making process. "I think what we have here is a rather unique tour," says Louis, a man you can tell is not one for bragging without reason. "We have a geode. We advertise it as the only geode of its kind, and we haven't been disputed yet." A geode is a rock with a cavity that's lined with crystals. But what's a geode doing on a winery tour? "My grandfather found it digging for water." The Crystal Cave has a deposit of celestite crystals, and a trip down into the cave is part of every tour.

"Anybody who goes on the tour gets a token for a glass of wine or grape juice," says Louis. You may also buy wine by the glass for $1 or an assorted cheese plate for $3.25. You may enjoy your wine and cheese outside in the wine garden or inside in the tasting room. The newly enlarged tasting room features wood carvings by the late island artist Bruno Weber.

Heineman's offers a full line of wines from the white Dry Catawba, and the dry red Claret, both of which are priced at $4.24, to the sweeter white Vidal Blanc, at $5.24, and the Pink Catawba, at $4.24. "My favorite wine is the Sauterne," says Louis. The Sauterne is a medium-dry white available for $4.24. The wines come from the thirty-five-acre vineyard on the island, where Gustav originally planted vines in 1885.

This island winery has character in the wine, the tour, and the family that runs it. Don't miss the boat—or in this case, the ferry.

DIRECTIONS: From the Ohio Turnpike, I-80, take exit 6 and drive north on Route 53 to Port Clinton. Port Clinton offers island service by Jet Express for passengers only. Those who wish to take their cars should continue on Route 53 to where it ends at Catawba Point. Miller auto/passenger ferries leave from this dock to Put-in-Bay. On the island follow the signs to the winery.

HOURS: Tours offered daily from May 5 to September 30 from 11 A.M. until 5 P.M. Tastings offered from April through October, Monday through Saturday from 10 A.M. until 10 P.M. and Sunday from noon to 7 P.M. Open year-round for sales.

EXTRAS: Winery sells wine, wine accessories, and cave-related items. Picnic facilities.

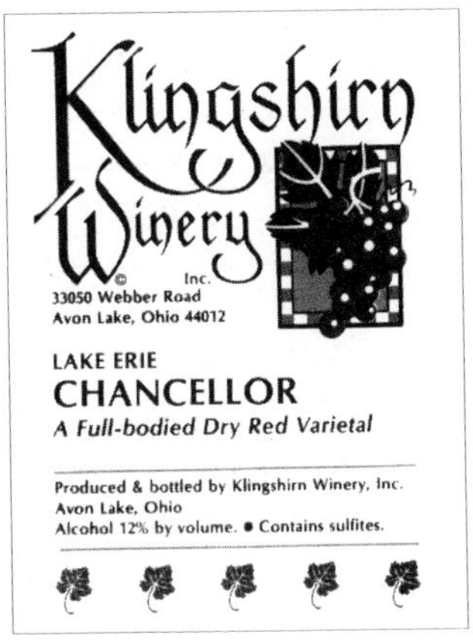

Klingshirn Winery

33050 Webber Road
Avon Lake, OH 44012; (216) 933–6666

"I'm not a salesman," says the relaxed Lee Klingshirn. "Good wine should sell itself. But it doesn't. People still look at the price tag." Even if you look at these price tags, you wouldn't be disappointed. Concord, Sweet Concord, and Vin Rosé sell for $3.43. Niagara, Dry and Haut Sauterne, and Dry and Pink Catawba sell for $3.64. Delaware, Golden Chablis, Blush, and Cherry sell for $4.11.

In 1991 Klingshirn introduced three types of champagnes—a semidry White Riesling, a slightly sweet Traditional Blend, and a nearly dry Contemporary Blend—all 1991 medal winners at the Ohio Wine Competition. These wines sell for $8.99 to $10.99.

You may order any of these wines with customized labels for special occasions. The winery will find a label that matches the occasion, such as a wedding, anniversary, or holiday, and will print your special message on it, such as "Happy Anniversary Dorie & Howard, 1951–1991." Labels cost $2.50 each, or less for an order of twenty or more, in addition to the cost of the wine.

Lee and his parents, Allan and Barbara, continue the family tradition of winemaking that began in 1935. "My grandfather, with two of his neighbors, bought a press, and they brought the grapes here. He started in the cellar of the house," says Lee. The grandfather, Albert, later bought his neighbors' share of the press.

He built a winery when neighbors kept bringing him work. "People brought him grapes and had him make their wine." The winery continued to grow. Albert's son, Allan, bought the business in 1955 and continued the growth. By 1978 the winery had expanded to four times its original size.

Lee, the third generation, makes the wine. Says Lee of his dad, "He's pretty much in charge of the pickers and getting the grapes in here. He leaves it all pretty much up to me, the winemaking." Lee studied winemaking at Ohio State University. He also studied in Europe. "I had an internship in Germany for four months, and I traveled Germany and France for months and stopped at the major research centers of each region."

Growing grapes in Ohio can be tough work with the weather and animals. "We bury the Riesling, lay them under the soil," which means taking the vines off the trellises and piling dirt over them so the earth can keep them protected from the cold. This is the work that Lee loves. "Working in the vineyard, the peace and quiet."

When you visit the tasting room, if there aren't too many customers, one of the Klingshirns will take you on a fifteen-minute tour of the winery. The winery gets busy during harvest, so you may want to call and check to see if tours are possible the day you're interested. You're always welcome to taste the wine.

You also get a taste of something else. "Nobody can leave without trying some grape juice. We probably sell as much grape juice as we do wine. We go through 7,000 gallons a year," says Lee. The winery usually sells about 10,000 gallons of wine a year. But with grape juice and wine sales combined, Lee doesn't see Klingshirn as a threat to Gallo. "We do enough to pay the bills and pay the mortgages."

DIRECTIONS: From Interstate 90, take exit 153. Drive north on Route 83 for 1.8 miles to Webber Road. Turn left; winery is .7 mile on the right.

HOURS: Tastings available Monday through Saturday from 10 A.M. until 6 P.M. Tours available during regular hours when time permits. Winery closed Sundays, New Year's Day, Easter, Thanksgiving, and Christmas.

EXTRAS: Winery sells wine.

Markko Vineyard

South Ridge Road
Conneaut, OH 44030; (216) 593–3197

"We don't tend bar," says Arnulf Esterer, winemaker at Markko. Nor do they put on a big show. But if you're looking for a place where people take their product seriously, you'll find it here. Markko is Ohio's pioneer of the European vinifera.

"We have their [French] scions. We have their barrels. But we're not going to end up there. We don't want to copy," says Arnulf, who with Thomas Hubbard, owns the winery.

One way Arnulf creates his own wine is by using some Ohio white-oak barrels. He was on a tour of the Mondavi winery in California and found that it used Ohio barrels. He thought if it was good enough for Mondavi, it was good enough for him. Arnulf cut some oak trees from his property and had barrels made from them.

You can see the barrels on the twenty-minute tour. "You may have to root us out of the vineyards or the cellar," Arnulf says, as no tour guide is standing by. "But we'll have the cheese and bread ready, the glasses ready. They can taste all they want. We just want people to drop in." When you drop in during the summer, bring a picnic. Or if you drop in during the winter, bring your skis. "We have cross-country ski trails for people who like to ski," says Linda Frisbee, the winery's only full-time employee. "We're open all year-round."

During the winter you can see how Markko protects its vinifera. "We mound up over the graft [on the vine] to protect it," says Linda. By burying the grafted vines, Arnulf has been able to grow vinifera where other vineyards have failed. "In the spring he [Arnulf] takes a tractor with a grape hoe [to uncover the plants]. The top of it takes its chances," Linda says.

Pinot Noir is one of the wines produced at Markko. Others include the 1988 Chardonnay for $16, the 1990 sweet late-harvest Riesling for $18, and the 1988 Cabernet Sauvignon for $13. Markko sells nonvintage bottles of Riesling, Chardonnay, and Cabernet Sauvignon for $5 to $7.50. Arnulf also makes a blend, Underridge White, for $4.

Grab your skis or pack a picnic and head for bucolic Markko.

DIRECTIONS: From Cleveland, take exit 235 off of Interstate 90. Drive north on Route 193. Turn right on Main Street, which runs into South Ridge. Winery is on the left. From Interstate 90, coming from the east, take exit 241. Drive south on Highway 7 for .5 mile.

Turn right onto South Ridge Road for 3 miles. The winery is ahead on the right.

HOURS: Tours and tastings available Monday through Saturday from 11 A.M. until 6 P.M. Winery closed Sundays, New Year's Day, Easter, Thanksgiving, and Christmas.

EXTRAS: Winery sells wine. Picnic facilities.

Meier's Wine Cellars

6955 Plainfield Pike
Cincinnati, OH 45236; (513) 891–2900 or (800) 346–2941

When you pull into the parking lot of Meier's Wine Cellars, you'll notice a round piece of cement. Unfortunately, that's all that's left of the jug-shaped structure where John C. Meier sold his grape juice in 1906. But you can see a picture of the jug on a Meier's tour.

The history of the oldest and largest winery in Ohio began in the 1850s when John M. Meier came to the United States and planted a vineyard with vines from his native Bavaria on his farm in the Cincinnati area. After having problems with the vines due to the cold winters and spring humidity, John M. and his son, John C., replanted the vineyard with the American Catawba variety.

The Catawba grape juice has been a large part of Meier's success. The Meiers found a process that would keep the juice fresh long after it had been pressed from the grapes. The juice became so popular that the winery moved in 1906 from its first location in Kenwood to its present location in Silverton, a suburb of Cincinnati, because the business needed a railroad siding to ship its goods.

As Cincinnati grew and took the land around the winery, Meier found he needed to go elsewhere for vineyard acreage. In the 1930s he bought the Isle St. George, Ohio, 1.5 miles from the Canadian border in western Lake Erie. Vines had been growing on the island since 1845. The 660-acre island is 1.5 miles wide and less than that in length. You'll receive a placemat-sized photograph of the island at the beginning of the 30-minute tour.

You'll also receive a description of the production process and a chance to walk through the cask area that was part of the original cellars. After the tour you'll be offered samples at the tasting bar, which is made of lucite filled with pebbles from Isle St. George. You can taste six or seven wines such as the Harvest Blush, which sells for $3.49; Peach La Brusca, $2.29; or Haut Sauterne, $2.99. The No. 44 Cream Sherry, $4.99, has brought Meier's much recognition, winning top medals over Bristol Cream.

Meier's makes a line of varietals—Cabernet Sauvignon, Chardonnay, Gewürztraminer, and Johannisberg Riesling, for example—named in honor of Nicholas Longworth, Cincinnati's famed pre-Civil War wine pioneer.

After the tours you'll be poured a selection of mostly sweet wines. You can skip the free tasting that comes with the tour, and for $1, pick the wines you want to taste and get a souvenir glass as well. Or you can do both tastings if you're interested.

DIRECTIONS: From Dayton, drive south on Interstate 75 and exit at 10B. Drive east 3.5 miles on Galbraith Road to Plainfield. Turn right and drive .9 mile; winery is on the right. From Cincinnati, drive north on Interstate 71 to exit 12. Drive west on Montgomery Road to Plainfield. Turn right and drive north; winery is on the left.

HOURS: Tastings available Monday through Saturday from 9 A.M. until 5 P.M. Tours available from June 1 to October 31, Monday through Saturday on the hour from 10 A.M. until 3 P.M. and other times by appointment. Winery closed Sundays, New Year's Day, Thanksgiving, and Christmas.

EXTRAS: Winery sells wine, wine accessories, and gifts.

OKLAHOMA

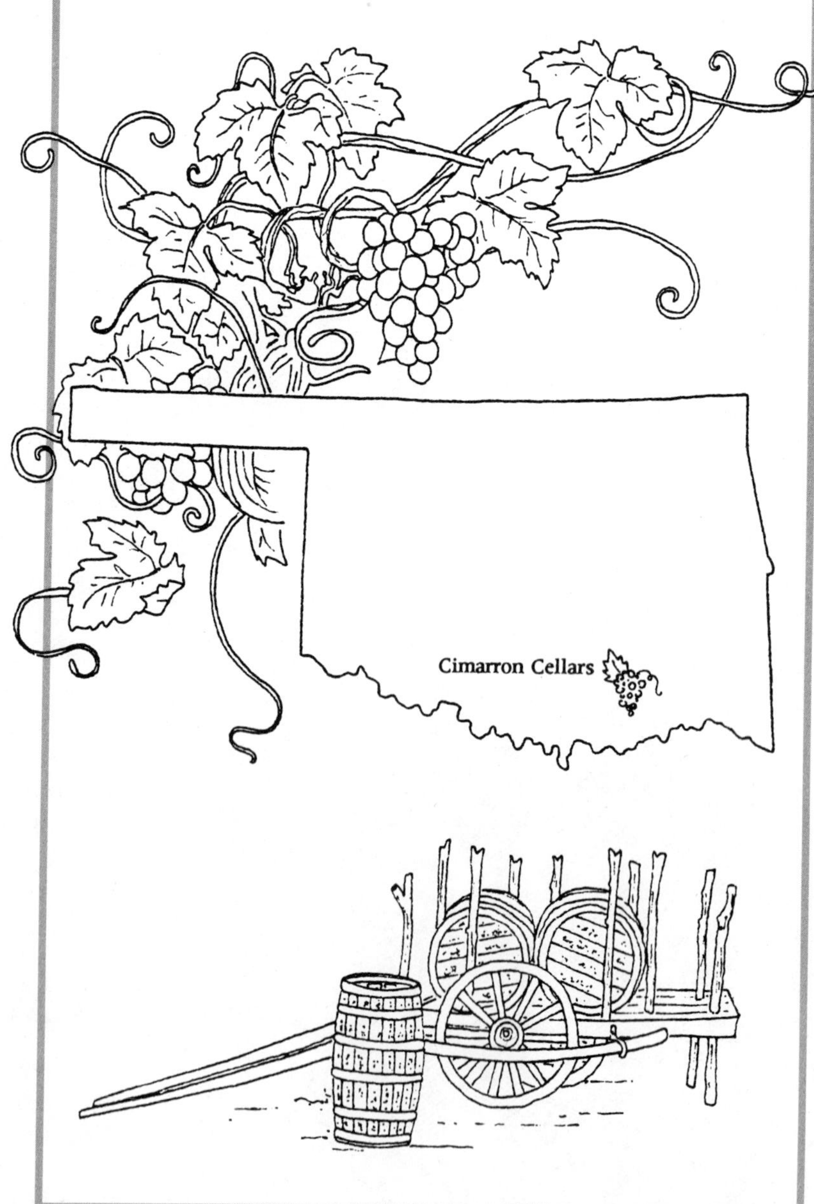

Oklahoma is known for having oil gushing from the ground, but the state's winemakers are trying to get the wine gushing. It's not easy in a state that took until 1986 to make it legal for people to sell wine at their wineries—and then only if the wine is made from Oklahoma fruit. Perhaps things will change in Oklahoma since the law passed, but at present only one winery operates in the state.

Dwayne and Suzé Pool, in Caney, licensed their winery in 1983. "I guess we're the only winery, and that's not really an advantage," says Dwayne. "One advantage is the land prices."

For several years they could show people around the operation but not sell wine. "People would drive all the way out here and we couldn't sell," says Suzé. Nor could the Pools tell people where to buy their wine in stores. The Pools worked through distributors, and according to Dwayne, the distributors didn't keep track of what stores were carrying their Cimarron Cellars wine. But the Pools hope all that has changed.

Although Cimarron is the only winery, it's not the only vineyard. "There's quite a few grape growers in Oklahoma," says Dwayne. Vineyards in the state are planted in both vinifera and French-American hybrids. Some of the vinifera are Zinfandel, Petit Sirah, Cabernet Sauvignon, and French Colombard. Seyval Blanc, Vidal Blanc, and Maréchal Foch are among the hybrids. The growers sell their grapes to home winemakers or to wineries out of state.

When the Pools bought the land in Caney, the vineyard was already there. During the early 1970s, the federal Office of Economic Opportunity purchased the land and planted grapes. In an effort to offer homes and jobs for the unemployed, mobile homes were put on the property with the idea that people would live in them and work in the vineyard. Ten families moved in, and in less than a year, nine were gone.

The project failed, and the vineyard fell into disrepair. Several owners held the land but didn't take advantage of the vines. The Pools hope to develop not only the vineyard but also a wine industry in Oklahoma. "I think they [the state] need to get behind it. We couldn't even advertise until July 1986," says Dwayne.

Their battle will be uphill. Although the state has allowed them to sell wine, they are not legally allowed to offer tastings at the winery. (In the spring of 1992 a bill was pending allowing samples at the winery.) Without supportive legislation Oklahoma's one commercial vineyard and winery could be killed by the same state that planted it.

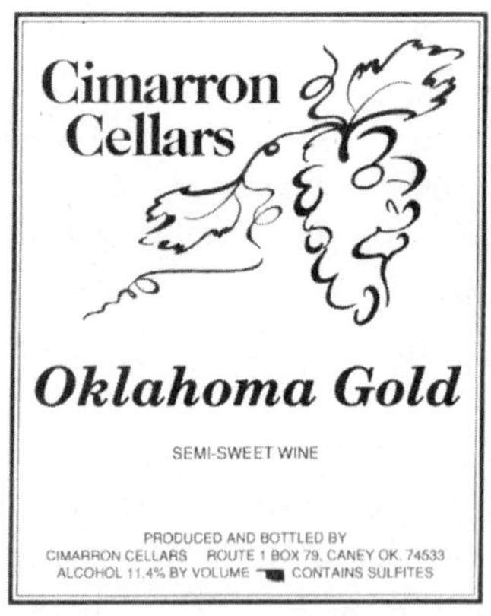

Cimarron Cellars

Route 1, Box 79
Caney, OK 74533; (405) 889–6312 or (405) 889–5997

"You know Oklahoma was dry until 1959," says Suzé Pool, who owns Cimmaron Cellars with her husband, Dwayne. He adds, "We're a dry county. You just can't buy liquor by the drink." Being in a dry county, the Pools can't legally offer samples at their tasting room. But problems are nothing new to them.

When the Pools bought the land, they had no immediate plans to open a winery. "We knew this vineyard was here. We were told that Arkansas would buy all of our grapes, but it wasn't so," says Suzé. Consequently, in 1979 the Pools had tons of grapes and no one to buy them. Dwayne scrambled around the country and found a solution. "When Arkansas didn't buy the grapes, we found a winery in Texas that would," says Suzé.

Life smoothed out for the Pools for several years until the next bump came. Suzé says, "In 1983 the Texas winery told us the day before harvest that he wanted only the white grapes." So with tons of red grapes nobody wanted, the Pools decided to start the winery they had thought was years in the future. They released their first bottle in 1984. The winery produced 1,200 gallons the first year, and now production has increased to 12,500 gallons.

The state had originally planted approximately twenty acres, and the Pools have added twenty more. "We're going to eliminate the varieties that don't do real well," says Dwayne. "In 1987 I got rid of the Barbera. I didn't really get anything from them. If the vinifera do as well on the spray schedule, we might try and graft more on."

The spray schedule is necessary due to the state's humid climate. When asked what the biggest problem is, Dwayne tells you, "It's really the weather." The humidity promotes a fungus called black rot. "In growing season we have to fight the fungus; it's like Ohio," says Dwayne of a disease that almost wiped out Ohio's wine industry. "We used an Ohio spray schedule. So now maybe we're on the right track."

Besides losing crops to the fungus, the Pools sometimes couldn't even get the plants to bear fruit. "We were only getting vinifera every three years, but maybe now we can grow them," says Dwayne. The spraying made the difference. That is the only major problem in the vineyard. "The raccoons do real well with the row by the woods, and we just expect that."

What grapes the fungus and raccoons don't get are transformed into wines such as the Vin D'OK, a mix of Ruby Cabernet and Cabernet Sauvignon, or the Oklahoma Gold, Blanc D'Blanc, and Sweet Suzé. A big seller for the winery is the Sooner Red. "They sell real good, and I think it's because of their name," says Suzé of a wine that uses the nickname of Oklahoma residents. Wines sell for $4.25 to $7.25.

The Pools weren't happy just sitting back and selling wine. Dwayne believed tourists driving by on the highway missed the turn off to the winery. "What we really wanted to do is get on the highway," says Dwayne. And in 1988 they did just that. The Pools opened a salesroom on Highway 69, just north of the road where you would turn off to go to the winery. Salesroom hours are Monday through Friday from noon to 5 P.M. and Saturday from 10 A.M. until 5 P.M.

When you don't miss the turnoff, you'll find that Cimarron Cellars is not only a winery but a showcase for Suzé's talents as well. "Dwayne makes the wreaths and I decorate them," says Suzé of the grapevine wreaths she sells. The tasting room is also filled with her oil and watercolor paintings.

Stop by the winery for the twenty-minute tour and buy some wine. You wouldn't want to be like most of the Pools' customers. Suzé says, "They say, `I've been traveling this road for ten years, and I finally decided to stop. I've always been meaning to stop.'"

Don't wait ten years to give the Pools the opportunity to show you firsthand the spunk that has gotten them this far. You won't be disappointed, and the Pools deserve a break.

DIRECTIONS: From the Atoka southern city limits, drive south on Highway 69 for 12 miles. Turn left at the Cimarron Cellars sign. Drive 4.7 miles east; winery entrance is on the left.

HOURS: Tours daily, except Sunday, from noon until 5 P.M. Closed Sundays, New Year's Day, Easter, Thanksgiving, and Christmas.

EXTRAS: Wine, wine accessories, wreaths, and paintings for sale. Picnic facilities.

OREGON

Amity Vineyards

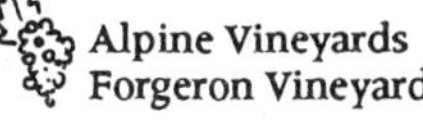

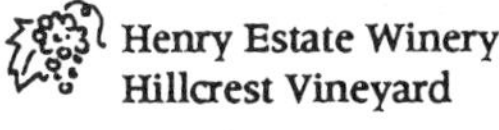

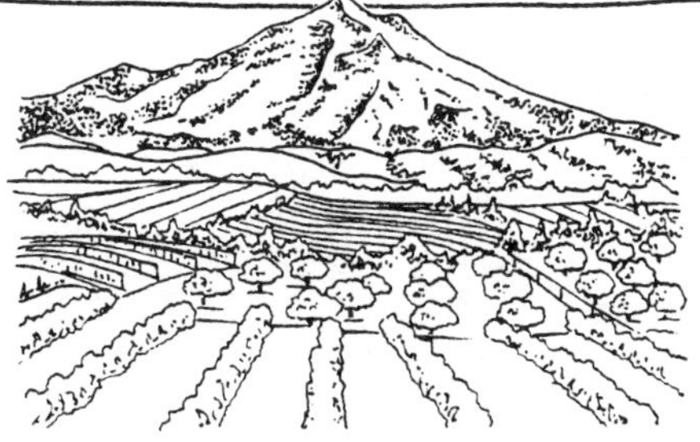

In 1847, Henderson Luelling came to the area now known as Oregon. He carried grape vines with him and experimented with different varieties in the climate southeast of Portland. Also important in early experimentation was A. R. Shipley, who worked with more than forty varieties of grapes from the East Coast and Europe.

In general the early work did not go well. It wasn't until 1890 that the first winery opened. Adolf Doerner built a winery near Roseburg. The few others that opened were closed by Prohibition. After the repeal of Prohibition, Oregon winemaking took off, and by 1937 twenty-eight wineries were producing wine. But within the next thirty years only a few remained.

Richard Sommer, however, brought winemaking back to the town of the original winery. He was attracted by the area's climate and soil. Sommer opened Hillcrest Winery in 1963. After much experimenting he met success with the White Riesling.

More vineyards opened and grape growers found more varieties suited to the state's climate, such as the Pinot Noir. Vineyards spread across the state in the valleys of the Columbia, Umpqua, Willamette, Rogue, and Wala Wala rivers. The wines produced from these vineyards have won national and international competitions and have helped Oregon become the hottest new wine-growing state.

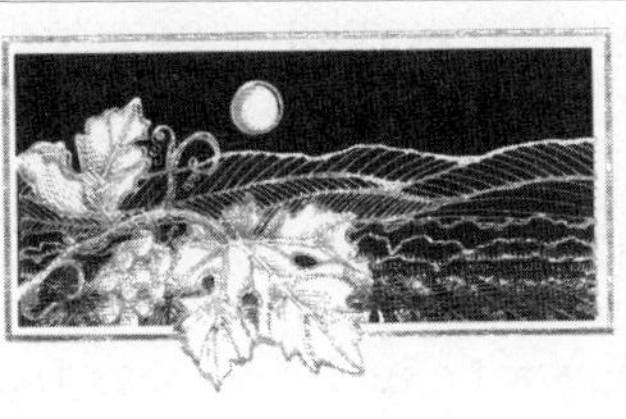

ALPINE VINEYARDS

ESTATE BOTTLED

Willamette Valley

1988 PINOT NOIR

Alpine Vineyards

25904 Green Peak Road
Alpine, OR 97456; (503) 424–5851

"We spent the summer driving around the United States deciding where to plant the vineyards. Then we moved to Eugene and started looking for winery property. In terms of the soil and the exposure, it seemed perfect," says Christine Jepsen, who with her husband, Dan, owns the land that holds Alpine Vineyards.

Alpine Vineyards rests in the foothills of Oregon's Coast Range, overlooking the Willamette Valley. Grapes do well in the cool growing season. "From a quality standpoint, this is the perfect place," says Dan. Christine agrees. "I'm from California originally. We think the longer, cooler growing season is ideally suited to certain varieties."

She says they were surprised at the beginning by the lack of insects or other problems. The only minor problem is the deer. Watch out when you drive to the winery—the deer are everywhere! "They don't eat the grapes, just the leaves," says Dan.

The land had previously been used for pastureland for sheep. "The only other thing it would be good for is Christmas trees," says Dan.

"It takes $10,000 to develop vineyard land in Oregon. We economized by doing a good portion of the work ourselves," says Dan. Christine and Dan first planted ten acres of vines in 1976 and by 1988 had doubled their acreage. Their production has grown from 2,500 gallons to 14,000.

"I started making wine in 1966," says Dan. "I learned the art of winemaking as a home winemaker." Christine says initially Dan's chemistry background helped with making the wine. Dan is a doctor at the University of Oregon in Eugene, and Christine is a part-time nurse.

Dan and Christine have moved their winemaking from the house to three air-conditioned buildings. "The cold fermentation retains the delicate flavor of the Riesling grape," says Christine. After fermentation the wine that is aged will go into one of the four types of French oak barrels that the Jepsens use. They will oak age their Chardonnays from three to six months and the Pinot Noir and Cabernet Sauvignon from one year to a year and a half.

You can see all the buildings, the oak barrels, and equipment on the Alpine tour. The tour length depends on your interest and time constraints. You'll probably want to spend most of your time in the tasting room, which affords a spectacular view of the vineyards and tree-covered hills beyond. Or do the tour and tasting, buy a bottle, and sit outside the tasting room while you enjoy the view and the wine.

The Jepsens also produce Riesling, Gewürztraminer, and a White Cabernet. All of Alpine's varieties have won prestigious awards. Wines sell for $6.95 to $16.95. "They [Cabernet Sauvignon, Pinot Noir, Chardonnay] are like my children," says Dan.

All the Jepsens wines are estate bottled—that is, the grapes are grown at the winery. "We are committed to being estate bottled. That's very important from a quality standpoint," says Dan.

The quality must be showing: Alpine Vineyards' wine is sold from California to Michigan, from England to Japan, and more people are asking for it all the time.

DIRECTIONS: From Eugene, drive north on Highway 99W for 23 miles. At the northern end of Monroe, turn west on the road to

Alpine. Drive 5.4 miles to the west side of Alpine and turn right onto Green Peak Road. After 1.5 miles you'll see the winery on your left.

HOURS: Tours and tastings available from June 15 through September 15, daily from noon until 5 P.M, and September 15 through June 15, weekends from noon until 5 P.M. Closed January.

EXTRAS: Wine and wine accessories sold. Picnic facilities.

Amity Vineyards

18150 Amity Vineyards Road SE
Amity, OR 97101; (503) 835–2362

"We're Pinot people," says Myron Redford, winemaker at Amity Vineyards. "It's hard to make, but when you hit it, there's no other wine that comes close."

Amity has "hit it" numerous times. In 1981 *Vintage Magazine* rated the winery's 1978 Winemaker's Reserve Pinot Noir as one of the eleven best in the United States. And in 1986 Amity became the only winery to win two of the four awards in a single category—second and fourth in the varietal Pinot Noir category in *Wine & Spirits* magazine's Gold Medal Tasteoff.

The winery produces two styles of Pinot Noir. One is barrel aged in oak like a Burgundy. The other, a nouveau, is a much lighter wine with a fruity taste like a Beaujolais.

As you'll hear throughout Oregon, this is the state for Pinot Noir. Besides the cool, long growing season, the rolling hills of the Willamette Valley provide just the slopes that this European varietal loves. You'll also hear quite often that Oregon is situated at approximately the same latitude as Burgundy, France—*the* place to grow Pinot Noir.

But the Pinot Noir isn't the only fine wine Myron Redford, Amity's owner, makes. He also produces Dry Riesling, Chardonnay, and Dry Gewürztraminer.

Since 1974 Myron has planted fifteen acres of vines. In 1991 he produced roughly 9,000 cases. The wine sells for $6 to $36 a bottle. You can taste the wine from a tasting table not 3 feet from a fermentation tank. What better way to watch the workings of the winery?

"People love the tasting room being in the working winery," says Myron. "We let them see the complexity and the magic of winemaking firsthand."

The different grape lots from the different area vineyards are kept separated throughout the fermentation and aging process, Myron explains. Then the people at Amity work at blending the different wines to come up with the right combination.

Each year the winery holds a summer solstice festival. "Our fermentation area becomes our fun area," says Myron. "After all, wine is for fun." There is music, food, wine tasting, and tours on Saturday and Sunday of the weekend closest to the solstice. It begins at 11 A.M. and runs until dark. Myron makes a Solstice Blanc in honor of the solstice, the longest day of the year and the official beginning of summer. Call ahead for the dates.

Whether in the summer or the winter, Amity is a place to visit. "The winemakers are gaining experience," says Myron. "I think the bottom line is that we're what Sonoma and Napa were ten to fifteen years ago."

Hurry and visit, before it gets crowded.

DIRECTIONS: From Salem, drive west on Highway 22 until you reach Highway 99W. Then drive north to Amity. On the north end of Amity, turn right, east, on Rice Lane. After .4 mile turn left onto Amity Vineyards Road. The winery is on the right after .6 mile.

HOURS: Tours and tastings available daily June through November from noon to 5 P.M. December through May the winery is open noon to 5 P.M. on weekends only. Closed December 24 through January 31. Appreciates a call for appointment for tours.

EXTRAS: Wine sold at winery.

Forgeron Vineyard

89697 Sheffler Road
Elmira, OR 97437; (503) 935–1117

"We try to make it a people place," says Linda Smith, co-owner with her husband, George Lee, of Forgeron Vineyard. They've succeeded. You can picnic under the grand fir trees or in the colorfully landscaped gardens. Sit beside a fountain or under the shade of an umbrella, or walk through the vineyards, up the hill behind the winery, for a panoramic view of the shining Fern Ridge reservoir. On a clear day you can see the snow-capped mountains of the Cascades.

"Each winery takes on the personality of the owners, as does the wine," says Linda. The personality of their wine is distinctly French. For some time Lee lived with a French wine-growing family. Even the winery's name reflects the French influence. The family's name is Smith, and *forgeron* is French for blacksmith.

"A lot of his original plants came from France," says Laura Magnuson, who works in the tasting room and directs tours. "Europeans tend to be interested in our wines because we tend to do a European-style wine." The Smiths also produce wines with delicate natures that complement food.

The Cabernet Sauvignon, for example, made in the French Bordeaux tradition, will add to any meal of roast beef or wild game. The Pinot Noir, in a dry Burgundy style, complements meals of lamb or roast pork. While the Riesling, a semisweet German-style wine, would serve you well before dinner or with chicken or shellfish. And the Rosé of Pinot Noir "was developed to complement barbequed salmon on mesquite wood. We also found this wine perfect with glazed ham and fowl stuffed with sage," say the Smiths.

They also produce a Chardonnay, Müller-Thurgau, Pinot Gris, and a Chenin Blanc. The wines sell for $5 to $20, and you may taste most in the tasting room. Each wine is explained and discussed.

"With the Chenin Blanc, the sugar level varies each year," says Laura as she pours the wine. "The Pinot spends one and a half to two years in oak." While serving the Müller-Thurgau, she explains, "The Müller name comes from the doctor who created this type of grape, and the Thurgau means spicy." Laura works four days a week at the winery, and when he's not busy making wine, Joe Goulart, the cellar master, conducts the tours and tastings.

You may receive a fifteen- to twenty-minute tour if you just stop by, or if you call in advance, you may receive a more formal, one-hour tour. The tour takes you past the water-cooled fermentation tanks, stainless-steel barrels, hand-made French oak barrels, machine-made American oak barrels, and the rest of the equipment.

The tour includes information about grape growing in Oregon.

"At the end of June the vines bud out. We harvest mid-October," says Laura. "We're dealing with a shorter growing season."

The Smiths deal with twenty-three acres of vines and 23,000 gallons of wine each year. They began in 1972 with five acres of grapes. The wines they produce are all award winners, from a gold for their 1983 Pinot Noir from the Enological Society of Seattle to the gold for their Riesling from the Indiana State Fair.

Whether it be the wine they create or the colorfully landscaped winery they have arranged, as the motto on the Forgeron label says, "Good taste is our way of life."

DIRECTIONS: From Eugene, drive west approximately 15 miles on Highway 126 until the Elmira and Florence Junction. As you come into the town of Elmira, you'll see Warthen Road. Turn left and drive 1.5 miles to Sheffler Road. Turn right; after 1 mile winery is on the right.

HOURS: Tours and tastings available daily from noon until 5 P.M. from June through September. Tours and tastings available October through May on Saturday and Sunday from noon to 5 P.M. Closed the month of January, Thanksgiving, and Christmas.

EXTRAS: Winery sells wine, wine accessories, T-shirts, and Oregon food products. Picnic facilities.

Henry Estate Winery

P.O. Box 26
687 Hubbard Creek Road
Umpqua, OR 97486; (503) 459–5120

"We never have a bad year with Pinot Noir," says Marlene LoPera, who works part time at Henry Estate Winery. "This is his [owner Scott Henry's] baby. He puts more into a Pinot to get it exactly what he wants it to be. In 1987 the 1984 [Pinot Noir] won a double gold in San Francisco, and if you know the competition, you know that is very hard to do." More recently his 1987 Barrel Select Pinot Noir won a gold from the 1991 Atlanta Wine Summit and an honorable mention at the American Wine Competition in New York.

Credit also goes to the land Henry uses for his grapes. The area not only takes in the scenic beauty of the Umpqua River Valley but has fertile land, too. "His grapes are all here on bottomland," says Marlene. Henry has a remarkably high yield on the rich land. "He gets six tons per acre, and that's after they've been thinned."

The river creates a climate that's excellent for grapes. "Just one year we had a little frost damage, and that's been the only problem. We have a nice breeze with the river," says Marlene. She should know, since besides working at the winery, she has a vineyard of her own. She planted ten acres in 1982 and seven more in 1988. She says they grew so vigorously that by the third year she had a harvest that usually takes seven years to obtain.

"We sell our [Chardonnay] grapes to California," says Marlene. Quite a testimony to the quality of grapes being grown in the area. But no one rests on the laurels. "Scott does a little bit of experimenting—different fertilizer, different things—and then he weighs each row.

"We don't do anything to the grapes without his direction," says Marlene. She also sells grapes to Scott. "We have smaller, darker berries, and Scott likes those to add a little deeper color to the Pinot." Pinot Noir is the grape used to make the famous French Burgundies. Scott ages his Pinots for two years in small oak barrels.

His background is in aeronautical engineering. Although he was a home winemaker in Sacramento, California, it wasn't until he returned to his family farm in Umpqua in 1972 that he decided to grow grapes on a grand scale. For the first three years, Scott and his wife and co-owner, Sylvia, sold the grapes from their twelve acres to other wineries. But it wasn't long before the winemaking bug struck again. They built their own winery and crushed their first harvest in 1978, producing 6,000 gallons.

For $5.50 to $18 you may choose a bottle of Henry's Pinot Noir, Chardonnay, Gewürztraminer, Cabernet Sauvignon, Riesling, or two blended table wines.

The Henrys offer a twenty-minute tour of the facilities. You will see the cool "Red Room," where Pinot Noirs mellow in their oak casks, and the refrigerated "White Room," where the white wines are kept cool to retain their crisp, fruity flavors.

You may also skip the tour and get right to the tasting, or do both. The tasting usually lasts another twenty minutes, or longer if you have a question about a specific wine. The people at the winery are friendly and eager to help and share their knowledge. Don't expect to taste the wine and run. Enjoy learning what is special about a particular vintage, what awards it has won, or what wines to use to accompany certain foods.

Don't miss the leaf imprints in the cement steps as you enter the winery. Henry took different varieties of grape leaves and pressed them into the wet cement. The veins create delicate, lacy, intricate patterns. The effect is stunning.

DIRECTIONS: From Interstate 5, drive 13 miles north of Roseburg and turn off at exit 136. Drive west for .3 mile, until you reach Highway 9. Turn left and drive 7 miles. After the Umpqua River crossing, the winery is on the right.

HOURS: Tours and tastings from 11 A.M. until 5 P.M. daily. Closed major holidays.

EXTRAS: Winery sells wine and wine accessories. Picnic facilities.

Hillcrest Vineyard

240 Vineyard Lane
Roseburg, OR 97470; (800) 736–3709

Although pioneers began grape growing in Oregon more than a hundred years ago, it took Richard Sommer's 1961 planting of vines in the Umpqua Valley to spark a resurgence of wine growing in western Oregon.

"I guess I was exposed to wine while growing up in San Francisco," says Richard Sommer, owner of Hillcrest Vineyard. People said it was too wet in Oregon for grapes, but he remembered his grandfather's old gnarled grapevines near Ashland and thought if vines could survive there for forty years they would survive in other locations.

In 1958 Sommer embarked on the study of climates from northern California to Vancouver Island, British Columbia, to find a location suitable for Riesling. He thought the Roseburg area was neither too cold nor too hot. Further research with the Forest Service and the Soil Conservation Service gave him the perfect spot: fourteen acres on a sloping hill west of Roseburg. He planted his first five acres of vines in 1961 and has since expanded the vineyard to thirty-five acres.

"The vineyard has twenty acres of Riesling," says Sommer. "Five acres each of Cabernet Sauvignon and Pinot Noir and five acres of other varieties such as Chardonnay, Gewürztraminer, Sauvignon Blanc, and Semillon. We use all our own grapes plus some Pinot Noir, Cabernet Sauvignon, and Chardonnay from the neighbors' vineyards."

Each year Sommer makes approximately 2,000 cases of Riesling and 500 cases each of Cabernet Sauvignon, Chardonnay, and Pinot Noir. "We do a new style of fruitier red wine where we gently stem the berries with little bruising." Wines sell for $4 to $18 a bottle.

While you taste the wine, you receive a description on how it is made and the history of the grape in Oregon. In this way Hillcrest helps you to understand what you are drinking, and in fact, the tasting becomes more informative than the tour.

For many people this is all they want from a winery. As with many small wineries not geared to tourists, if there are other people tasting and not interested in a tour or production is going on in the back of the winery, you may not have the opportunity to tour the facilities. For great wine and a view of a historic vineyard that brought commercial winemaking back to Oregon, however, this is the place.

DIRECTIONS: In Roseburg on Interstate 5, take the Garden Valley exit west for approximately 2 miles. Turn left on Melrose Road and drive for 3 miles. Turn right on Cleveland Hill Road and continue until you reach Orchard Lane. Then turn left. Orchard Lane dead-ends into Elgarose. Turn right; winery is approximately 2 miles on the left on Vineyard Lane.

HOURS: Tours and tastings available daily from 11 A.M. until 5 P.M. Closed major holidays.

EXTRAS: Wine sold at the winery. Picnic facilities.

Wasson Brothers Winery

41901 Highway 26
Sandy, OR 97055; (503) 668–3124

"The town is pretty proud of the winery. We took them [competitors] to the cleaners at the [1990] Oregon State Fair. We won Best of Show seven of the last nine years," says Jim Wasson, co-owner of the Wasson Brothers Winery with his twin brother, John. "I think they got tired of us winning all the time and that's why they discontinued it in 1991." In 1991 the fair began awarding medals instead, and the Wassons won a gold for their strawberry wine and a silver for their 1988 Pinot Noir.

The brothers had been part-time farmers since the 1960s, and they branched into wine grapes in 1978. Then Jim and John became amateur winemakers. As their expertise and interest increased, Jim quit his work as a plumber and opened the winery. "There's not another winery around for sixty miles," says Jim.

The winery rests under the shadow of Mount Hood, which is the highest point in Oregon. The building, with its tall false facade and broad sidewalk in front, looks as if it has come out of the old Wild West. You can almost hear the spurs jangling.

You enter the working winery through a door at the end of the tasting bar. During the fifteen-minute tour Jim will show you the fermentation tanks and their cooling systems. "The first tank I can heat up when dealing with frozen berries," says Jim about fermentation tanks used generally only for cooling. The Wasson brothers produce loganberry, raspberry, blackberry, and rhubarb wines. When the price of strawberries and boysenberries isn't too high they'll make those wines also. The berry wines sell for $7.50. Pinot Noir, Gewürztraminer, and Chardonnay sell for $12.

Jim will also show you his Italian press, which holds up to a ton of grapes. "It's probably forty or fifty years old," says Jim. What's older is his labeling machine. "It's got a 1912 date on there. It never misses a beat." The machine came from a horseradish plant. "It's a god-awful-looking thing, but it works." So does Jim. "I can do [bottle daily] about 250 gallons myself, or if I can con my wife into coming up, we can do about 500 gallons."

One wall of the tasting room appears as if a grape exploded on it—everything is purple. There are purple Wasson T-shirts and a vine wreath decorated with flowers of different shades of purple. But most of the purple comes from the ribbons the Wassons have won for their wines.

In front of the "grape wall" is the wooden tasting bar. When you're not feeding crackers to Groaner, their Dalmatian, you may taste the wine. Jim's favorite is an early Muscat. "It's my favorite grape to grow. The vineyards smell just like wine and the grapes hang down like they should," says Jim. "When I'm sitting at home, I'll drink the Muscat."

If Muscat is not for you, don't worry. "The advantage for us is that we have something for everyone. Everyone leaves with a bottle."

DIRECTIONS: From Portland, drive east on Highway 26 for approximately 30 miles to Sandy. The winery is less than 1.5 miles east of Sandy, on the north side of the highway.

HOURS: Tours and tastings available daily from 9 A.M. until 5 P.M. Closed New Year's Day, Thanksgiving, and Christmas.

EXTRAS: Winery sells wine, wine accessories, wine- and beer-making supplies, and T-shirts.

PENNSYLVANIA

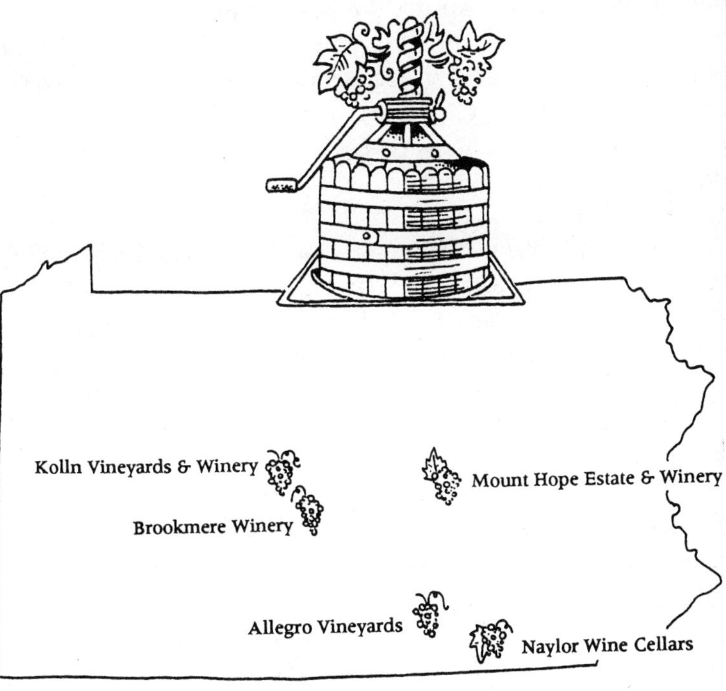

William Penn, who in 1683 was governor of the state, has been credited with planting the first European vinifera vines in Pennsylvania. He established a vineyard in the Philadelphia area. Others tried with varying degrees of success, but there were no commercial wineries until 1793, when Pierre Legaux opened his Pennsylvania Wine Company. Legaux's operation continued until the 1820s.

Grapes were also being grown in the northwest corner of the state. Vineyards planted with native American varieties began in the late 1700s. The vineyards expanded with the arrival of Charles Welch, whose company needed Concords for its grape juice. The vineyards in the area remain the largest plantings of grapes in the state. According to the Pennsylvania Wine Association, approximately 14,000 acres of grapes are planted in the state, most of which are used by Welch's.

Prohibition closed down most of the wineries, and even its repeal in the 1930s couldn't bring them back. By 1933 Pennsylvania had organized state stores as the only places alcohol could be sold. This policy served as a detriment to would-be winery owners, since sales at the winery are often the major part of their business.

In the mid-1960s several home winemakers banded together to change the state's laws. In 1967 the farm winery bill was introduced, and a year later it passed, allowing wineries to sell wine and give samples. With the passage of the Limited Winery Act, winemaking in Pennsylvania took off. Since then the number of wineries and the quality of wine have been rising each year—a return to the Pennsylvania tradition begun more than three hundred years ago.

Allegro Vineyards

R.D. 2, Box 64
Brogue, PA 17309; (717) 927–9148

In musical notation *allegro* refers to a fast tempo. "Fast vineyards is an appropriate name. Everything happens so fast in the business," says John Crouch, owner of Allegro Vineyards with his brother, Tim. "Tim and I were musicians before getting into the wine business. Then we were amateur winemakers.

"I majored in oboe. Tim played violin. He drives tractors now," says John. "I haven't played since I got in the wine business." John may not play, but you'll hear classical music in the tasting room. The music may not be for everyone, but neither is the winery. "I'm not really interested in giving tours. I'm interested in enlightening people about the good wines of the East," says John. If that's where your interests lie, this is the place. Allegro wines have consistently won high awards and are well thought of in and out of the state.

"We get medals, some really high medals, and consistent medal winners." John has his own ideas about why his wines win medals. "Part of the reason is the weather. The grapes have to suffer a little more, but they give a little more of themselves." Some of the wines include the French-American hybrids Seyval and Vidal and the vinifera Chardonnay and Cabernet Sauvignon. The wines sell for $5.95 to $15 and the Brut sparkling wine for $18.

You can see the wine being made during the fall months—but don't expect a tour then. "They can watch the operation and not have us talk," says John. "September and October we're usually so busy that we can't talk. It's really a stress period. The best time is in January. Call ahead."

The price for the tour is $2, refunded with a purchase. But if you truly want to learn more about wine, you probably won't have to pay. "If they're not interested in the wine, we charge," says John. You can tell he'd rather be drinking or making wine than taking a tour-bus group around. But it's just that kind of enthusiasm for wine that can make visiting this winery a delightful learning experience. They say it best for themselves at Allegro: "No gimmicks, no mansions, no fancy gardens, just really good wines. Come and taste Pennsylvania's best."

DIRECTIONS: From York, drive south on Interstate 83 to the 6E exit, Route 74. Drive 13.2 miles east on 74. Just as you come into the town of Brogue, turn right at the Post Office, onto Muddy Creek Road. This is an easy turn to miss. Then drive 2.1 miles to winery sign and turn left. After .7 mile, winery is on the left.

HOURS: Tours and tastings Wednesday through Sunday from noon to 5 P.M. Call for an appointment during winter months. Winery closed New Year's Day, Easter, Thanksgiving, and Christmas.

EXTRAS: Winery sells wine and wine accessories. Picnic facilities.

Brookmere Winery

R.D. 1, Box 53
Belleville, PA 17004; (717) 935–5380

The people who run Brookmere, owners Don and Susan Chapman and their salesroom manager, Linn Irvin, have as much character as their wine. And they make good wine. "I was in the forging business," says Don with a smirk. "Iron, not hot checks." Their car license plate, too, has a touch of whimsy: "FYNYNE"—or "fine wine."

"Tours are by chance, or by appointment," says Don. You'll want to make an appointment so as not to miss out. Don will give you a complete forty-five-minute tour of the winery and winemaking process if you're interested. Don says he'll make it shorter "if they're not really interested in listening to the details." He says many people just ask "Where's the wine?" And that's fine with him. Unfortunately, they're missing a great tour in a historic location.

"Brookmere was the name of this farm before it was converted to a winery. It means 'source of the brook,' and there's a brook out back. The barn is a 1866 vintage, the same as the house," says Don. The winery is located in a stone-and-wood barn built by James

Alexander, who in the late 1700s also had vineyards. Among the most beautiful aspects of the barn are the huge wooden beams. "All hand-hewn beams. They were mined from the mountains of Pennsylvania." When Don shows you around, make sure you see the second-story railing. The rails are all ax handles. "All made from hickory. They were rejects but good enough for this," says Don.

The cozy tasting room is on the ground floor of the barn, which is nestled into a hill. Wicker baskets, homey wood cupboards, and grapevines wrapped around wood pillars make for a comfortable country atmosphere perfect for wine tasting. The Chapmans make several blended wines such as Valley Mist and Shawnee Red. They also produce Chablis, Niagara, Riesling, Chardonnay, Chambourcin, and Cabernet Sauvignon. "We run the whole range from sweet Niagara to dry Cabernet Sauvignon," says Don. Most of the wines sell for $3.95 to $7.25. The Cabernets are more expensive and sell for $9 to $12 depending on the year.

After tasting the wine you may want to spend more time in the area. "I think the winery's biggest asset is its location," says Susan. "You step back in time. These Amish haven't been exploited like Lancaster," an area known for its Amish population. "We're close to a lot of state parks." And Don adds, "There's excellent hunting and fishing in the area."

Whether you spend an afternoon or a few days, you'll enjoy sampling some of the state's beauty and history.

DIRECTIONS: From Harrisburg, drive north on U.S. Highway 322 to Route 655. Drive west for 5.1 miles; winery is on your right.

HOURS: Tastings are available from 10 A.M. until 5 P.M., Monday through Saturday. Open Sundays from 1 P.M. until 4 P.M. Tours available by appointment; impromptu tours given if Don is at the winery. Winery closed New Year's Day, Easter, Thanksgiving, and Christmas.

EXTRAS: Winery sells wine and wine accessories.

Kolln Vineyards & Winery

3628 Buffalo Run Road
Bellefonte, PA 16823; (814) 355–4666

The Kolln Vineyards grew from a Christmas present. "I gave my husband a home winemaking kit for Christmas. It started as a hobby," says Martha Kolln, who owns the winery with her husband, John. "He was a professor at Penn State. He retired to go full time. It's small enough to have fun, and we're just going to stay small," says Martha, who teaches English at Penn State.

The winery is strictly a family operation, with John as the viticulturist, winemaker, sales manager, tour guide, and all-around handyman. Martha helps with sales, advertising, and record keeping. The rest of the family joins in at harvest time.

The Kollns bought their farm in 1971 and started to plant grapes when John's winemaking operation wouldn't fit in their basement anymore. In 1978 they added a winery building. "My husband built the building with our two sons," says Martha. They were traveling in Idaho and got the idea for the shape of the building while looking at potato cellars. "This is a variation of that theme." The winery is built into a hill so at the front you're on the ground floor but inside you climb to the second floor to exit on the ground floor in the back. The partially underground construction keeps the building cool in the summer.

The winery property is filled with beautiful tall trees and is graced by a gently flowing stream. You can drink wine or picnic by the stream. The Kollns have provided the tables, and they'll supply the cold wine. You can also picnic on the hillside overlooking the winery. The Kollns have both varietals and blends in dry and semisweet varieties. One of the most popular is the Seyval, a dry, crisp varietal. Other dry wines are a red blend and Steuben, a rosé. Among the semisweets are Niagara, Martha's Blush, and Semisweet Red. The wines sell for $4.70 to $5.25. They also make carbonated wines, which sell for $7.50. Recent additions are Cabernet for $10.50 and Riesling for $8.50.

Martha or John will guide you through the winery and explain the winemaking process. They also have hung photographs of grape varieties on the wall in the tasting room. If you haven't had the chance to walk through a vineyard in the fall, these photographs will show you what each grape variety looks like.

But if at all possible, visit the beautiful vineyards in weather suitable for strolling among the vines.

DIRECTIONS: From Interstate 80, take exit 23 to Bellefonte. Drive south 8 miles on Route 550. The winery will be on your left, 2 miles south of the Fillmore general store. From State College take U.S. Highway 322 west to Stevenson Road. Turn right and drive through Waddle to Route 550. Turn right; winery will be on the right after 1.5 miles.

HOURS: Tours and tastings available Saturday from 10 A.M. until 6 P.M. and Sunday from noon until 4 P.M. Winery closed Easter, Thanksgiving, and Christmas.

EXTRAS: Winery sells wine and wine accessories. Picnic facilities.

Mount Hope Estate & Winery

P.O. Box 685
Cornwall, PA 17016; (717) 665–7021

Mount Hope Estate & Winery offers you a tour of the estate, not the winery. Instead of damp musty cellars and fermentation tanks, you will be greeted by Egyptian marble fireplace mantels, brass and crystal chandeliers, hand-painted ceilings, and daisy-shaped decorations made from wood.

The Mount Hope Mansion was built in 1800 by Henry Grubb, a wealthy ironmaster. His great-granddaughter, Daisy Grubb, added twenty-two rooms to the original ten and lived in the mansion until she died in 1934.

Daisy redecorated the house in Victorian style, adding her personal touch—daisies—in interesting and unusual places. The tour guide points them out on the forty-five-minute tour through the house. Tours cost $4 and enable visitors to see about ten rooms of the house, such as the dining room, library, and Daisy's bedroom. The tour begins at the front of the house with a history of the family. Then the tour guide explains the changes Daisy made and why; for instance, she removed a mantel from a fireplace because she thought it looked too feminine. You'll see where Daisy hid her valuables and how, to the amazement of her friends, the servants magically appeared when she needed them. (There were secret buttons hidden under the rugs.)

After the tour you go to the billiards room for a wine tasting. The tour guide tells how to taste wine, by giving pointers on how to check for color and clarity, how to swirl the wine to release the bouquet, and how to roll the wine around your mouth to appreciate its flavor fully. The guide also talks about aging and storing wine. You're given four wines to taste of the more than twenty wines made.

The winery produces approximately 60,000 gallons a year. Licensed in 1973, Mount Hope is a joint venture with Charles J. Romito and Frank Mazza. The winery owns ten acres of grapes from which it produces its wine. Prices range from $5.95 to $12.95, but don't expect to taste the higher-priced wines. You'll be given tastes of its medium-dry Chablis, medium-sweet Sauterne, sweet Pink Catawba, or sweet Niagara.

Mount Hope is famous for its Renaissance Faire held weekends each year beginning at the end of June and running through the middle of October. The grounds are transformed to recreate a sixteenth-century Elizabethan country fair, complete with glassblowers, puppeteers, storytellers, and jousting armor-clad knights. The estate also has "A Charles Dickens Christmas Past" with candlelight tours. Actors assume the roles of such Dickens characters as Tiny Tim, Scrooge, and David Copperfield for performances staged in the house. The mansion is decorated with evergreen and candles. Festivities include caroling, sampling hot spiced wine, and lighting a yule log. Call the winery before you visit because times and dates for both events are subject to change.

Whether it's jousting or caroling or just wine, you'll find it at Mount Hope Estate & Winery.

DIRECTIONS: From Harrisburg, drive east on the Pennsylvania Turnpike and exit at Route 72. Drive south for less than one mile, and winery is on the left.

HOURS: Tours and tastings offered Monday through Saturday from

10 A.M until 6 P.M. and Sunday from noon to 6 P.M. Winery closed New Year's Day, Election Day, Thanksgiving, and Christmas.

EXTRAS: Gift shop sells wine and wine accessories.

Naylor Wine Cellars

R.D. #3, Box 424
Stewartstown, PA 17363; (717) 993–2431

"The bottom line is finding something you enjoy, and that's what's so neat about visiting a local winery," says Edward Potter, cellar master for Naylor Wine Cellars. "We readily send people to other wineries." Although they'll send you to another winery, you probably won't have to go farther than Naylor's for good wine and hospitality. They aim to please—whether it's a vineyard and winery tour or a tasting of their more than twenty wines. You'll probably end up like Edward when he came to work at Naylor: "I was hooked on Naylor wine."

You can tour the winery anytime and may receive a vineyard tour if you call in advance. If you don't call ahead, though, you can still see the vineyards. "We don't mind people walking in the vineyards. People will buy a bottle of chilled wine and have a nice time out there," says Edward. And if you don't want to sit among the vines, there's a gazebo for public use.

You're also invited to visit during harvest and watch the wine being made. "Our harvest runs from August to the middle of October. People are welcome to come out and see it. We do have a lot to offer," says Edward.

"We" includes owners Dick and Audrey Naylor. Dick is a former executive vice-president and general manager for a York corrugated box company. "He calls this his hobby getting out of hand," says Edward. Edward is a chef who has joined the Naylor family as both cellar master and consultant. "I was teaching culinary arts and decided to give this a try. I'm basically self-taught."

Naylor Wine Cellars produces approximately 5,000 gallons of wine a year. It makes a wine for every taste, from what Dick calls the "socially sweet" Niagara and Concord to the "deliciously dry" Seyval and Riesling. Prices range from $3.85 to $9.85.

Edward says Naylor works hard to produce high-quality reds. "We have a total of about sixty [oak] barrels. We're probably doing more than any other winery in the state. We talk about building complexities in reds." Some of Naylor's reds are Chambourcin, Cabernet Sauvignon, and Pinot Noir.

The winery offers three special programs in celebration of the grape. The Grape Blossom Festival has food, music, crafts, wine tastings, and hayrides. It runs both Saturday and Sunday the first full weekend in June from 11 A.M. to 7 P.M.

The second program, Harvest Festival, brings grape stompings to Naylor. It also has food, music, wagon rides, and wine tasting. Harvest Festival runs the first full weekend in October on both Saturday and Sunday from 11 A.M. to 6 P.M.

The third program offers picnics at the winery each Sunday in June, July, and August. A different buffet (along with wines, of course) is served each weekend. The buffet runs from 1 P.M. until 4 P.M, and reservations are required two days before the Sunday picnic you wish to attend.

Naylor also has a Wine Shoppe in York north of the winery. Located in the pavilion at Whiteford Road and Route 24 and next to the York County Tourist Bureau, it offers samples of the Naylor wines. It's open Monday through Saturday from 10 A.M until 9 P.M. and Sunday from noon to 5 P.M.

DIRECTIONS: From York, drive south on Interstate 83 to exit 1. Drive east on U.S. Highway 851 for 4.6 miles to Route 24. Turn left on 24 and drive north for 2.2 miles. Winery is on the left.

HOURS: Tours and tastings offered Monday through Saturday from 11 A.M. until 6 P.M. and Sunday from noon to 5 P.M. Winery closed New Year's Day, Easter, Thanksgiving, and Christmas.

EXTRAS: Winery sells wine and wine accessories. Picnic facilities.

RHODE ISLAND

More than forty years after the repeal of Prohibition, wineries returned to Rhode Island. The state's first estate-bottled wine was released in 1975 by Sakonnet Vineyards in Little Compton. According to the winery, it was not only Rhode Island's first estate-bottled wine but the first in all of New England.

Jim and Lolly Mitchell began Sakonnet Vineyards on a former 120-acre potato farm. The Mitchells scored another first in 1982 when they produced the state's first champagne. When they retired in 1987, they sold the winery to Susan and Earl Samson.

With the passage of the state's farm winery law in 1977, others began to plant vineyards with plans to open wineries. Another boost came to Rhode Island winemakers in the mid-1980s when the federal government designated Rhode Island as a viticultural area, called Southeastern New England.

Vineyards have done well in Rhode Island with the Narragansett Bay and Atlantic Ocean to help moderate the climate. The winery owners like to compare Rhode Island's weather and climate to that of the Burgundy area in France.

Grape growers plant both vinifera and French-American hybrids. Some of the wines being produced in the state are Vidal Blanc, Chardonnay, Pinot Noir, Pinot Blanc, Riesling, and Gewürztraminer. The owners of the young vineyards are experimenting with new varieties each year with the hope of putting Rhode Island at the forefront of the New England wine industry.

Diamond Hill Vineyards

3145 Diamond Hill Road
Cumberland, RI 02864; (401) 333–2751

Diamond Hill has the perfect gift to impress your family and friends or to celebrate a special occasion. For the price of only one case of wine, you can have your own personalized labels put on the bottles—for no extra charge. Peter Berntson, owner of the Diamond Hill Vineyards with his wife, Clara, does the full-color artwork. It's not something a large commercial winery could do. "It can only be done by a little guy like me," says Peter.

If you'd prefer to buy by the bottle, but are still looking for some special gift, you may buy single all-occasion bottles that have a card that matches the label. The Berntsons work with a greeting card company and use some of the artwork from its cards on their labels. Some of the cards and labels available are "Babies are Special," "Happy Birthday," "Happy Thanksgiving," and "Thank You."

If you want something even more special, Clara will take the all-occasion-labeled wine and make a gift pack for you. "I bring in all New England products and put them together." They include such appetizing items as smoked meats, Vermont fruit jellies, honey, apple butter, and Diamond Hill wine.

The Berntsons offer a full range of fruit wines: peach, blueberry, three types of apple, and raspberry. They also produce Pinot Noir and Pinot Blanc from their four-and-a-half-acre vineyard. Nongrape wines sell in the $5 range. Pinot Blanc sells for $9, and the Pinot Noir's price varies with the vintage.

You can taste the wine in one of the front rooms in the unusual Berntson home. Peter says the house was built two hundred years ago as a Cape Cod. Then a different owner, about a hundred years ago, added Victorian touches. The tour begins outside the house with a brief history of both the house and the farm.

If Peter gives you the forty-five-minute tour, you'll see where his interests lie. The tour takes you out into the vineyard to learn how the Berntsons manage to grow Pinot Noir and Pinot Blanc and have it survive the cold Rhode Island winters. "We take dirt and cover this area where the buds are," says Peter pointing toward the bottom of the plant.

Birds also pose problems. "Once they [grapes] change from green to black [mid-August], the birds see them. They'll peck at them and wipe you out in a day or two," says Peter. To protect the grapes he covers them with black plastic netting.

Next you visit the fermentation area. "We built these tanks for Pinot," says Peter of the tanks with cone-shaped tops. While most wineries move the wine through hoses from tank to tank, Peter just moves the tank. With the use of an overhead crane, Peter can drop the tank under the press to catch the juice as it is squeezed from the grapes, or raise the tank so gravity moves the wine into the bottler. "I knew I wanted to use gravity with the Pinot." He tries not to move the wine too much or filter it too often. "We don't like to beat our wine," says Peter.

When you return to the tasting room in the 200-year-old house and taste the wine, you'll think nothing can beat it.

DIRECTIONS: From Interstate 295, take exit 11. Drive north on Route 114 for 1.8 miles; winery is on the right.

HOURS: Tastings offered Wednesday through Monday from noon until 5 P.M. During November and December hours extend until 6 P.M. Tours offered Sundays from noon until 5 P.M. Winery closed Tuesdays, New Year's Day, Easter, Thanksgiving, and Christmas.

EXTRAS: Winery sells wine and wine accessories. Picnic facilities.

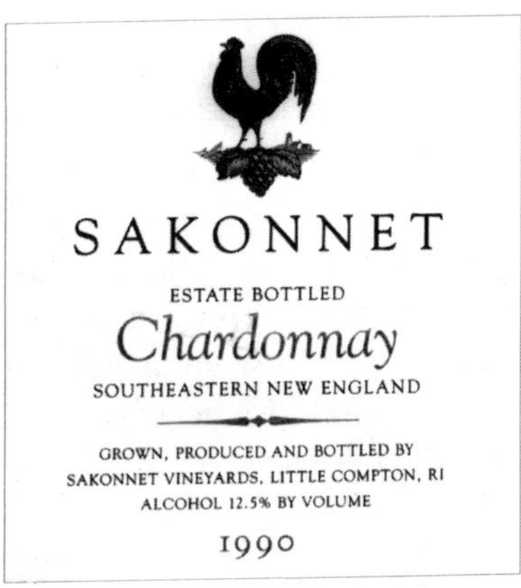

Sakonnet Vineyards

P.O. Box 197
162 West Main Road
Little Compton, RI 02837; (401) 635–8486

Sakonnet Vineyards is one of the largest wineries in New England. It began in 1975 when Jim and Lolly Mitchell bought an old potato farm and planted vines. Initially the vineyard grew mostly white varieties of French hybrids. Through years of experiments the Mitchells realized the climate was more appropriate for varieties of vinifera such as Chardonnay, Pinot Noir, and Riesling.

After years of successful work, the Mitchells decided to retire and sell the winery. In 1987 Earl Samson, a New York investment banker, and his wife, Susan, an actress and Broadway producer, bought the winery. Earl was a founder of Landmark Vineyards in Sonoma, California, in 1973.

They had spent summers in Rhode Island before deciding to make their stay year-round and return to the wine business. In 1987 they produced 40,000 gallons at Sakonnet. They hope to triple production in the near future.

The winery offers a twenty-five-minute, three-part tour. The tour takes you through a gray weathered-looking building and through the production process. If you're there in the late fall when wine is fermenting, you may have a chance to climb a ladder to the top of a fermentation tank and watch the mixture bubbling as carbon dioxide escapes. The winery also offers a twelve-minute presentation, which is available on nontour days as well. It shows the vineyard throughout the year and describes the winemaking process.

For those who've never had a chance to walk among the vines, there's a self-guided "vineyard walk." From the tasting room you pick up a sheet that shows where each grape is planted. During the preharvest days you may pick a grape. It's a wonderful experience to taste the grapes and then move into the tasting room to see what types of wine come from them.

The Sakonnet wines have been a success ever since the first bottle was released. Most years the winery sells out of its wine even before the next year's wine is available.

Sakonnet offers tastes of five or more wines, depending on what's available. The Chardonnay sells for $12.50; Gewürztraminer for $11.50; and the Rhode Island Red for $8.25. Because of the pride of the area from which the wine comes, many of the labels feature a Rhode Island Red Rooster. The Rhode Island Red is a hybrid chicken developed in Little Compton more than 130 years ago.

One of the wines, called Eye of the Storm, celebrates a disaster that was narrowly averted. The vines trembled on their trellises in 1985 as Hurricane Gloria raged up the East Coast and headed for Rhode Island. At the last minute the hurricane turned and hit Connecticut instead. The harvest was saved. Sakonnet celebrated by creating a blend. The label features a satellite picture of the eye of the hurricane. The wine became so popular, it continues as a tradition. With luck like that how can the winery not continue to prosper?

DIRECTIONS: Driving south on Route 77, you'll see the winery on your left 2.9 miles past the traffic light at Tiverton Four Corners.

HOURS: Tastings available daily from May 1 to December 31 from 10 A.M. until 6 P.M. From January 1 to April 30, daily from noon to 5 P.M. Guided tours offered on Wednesday through Sunday from May 1 to December 31. Winery closed New Year's Day, Easter, Thanksgiving, and Christmas.

EXTRAS: Winery sells wine and wine accessories. Picnic facilities available.

SOUTH CAROLINA

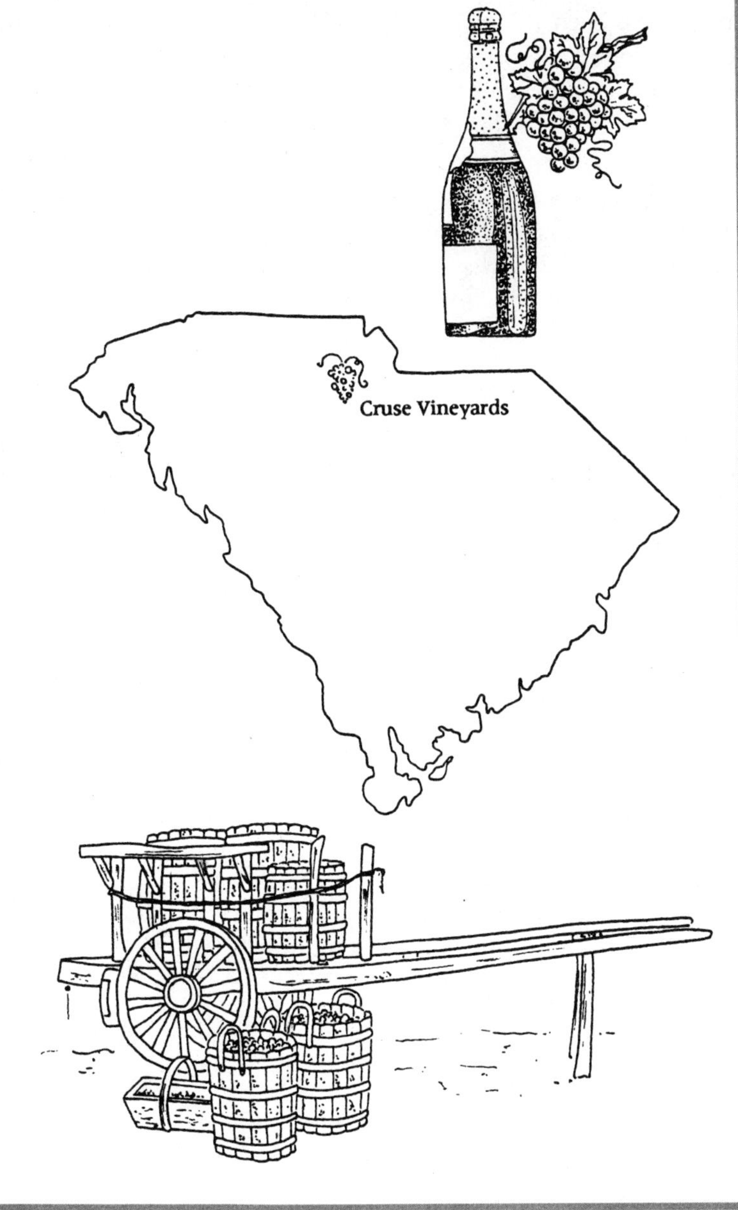

If it weren't for a £100,000 bribe in the late 1700s, South Carolina could have been the Napa Valley of the East Coast. According to research done by James P. Truluck, Jr., who previously owned a winery in the state, French Protestants seeking religious freedom came to South Carolina in the late 1600s and early 1700s. In 1764 the British government gave a group of these French 30,000 acres to create a wine industry in the upper Savanna River area. Louis de Saint Pierre came to the area, called New Bordeaux, to supervise the planting of the vines. His 1772 work called *The Art of Planting & Cultivating the Vine and Also of Making, Finishing & Preserving Wine* is considered to be one of the first books on American winemaking.

Saint Pierre's vineyards did well, and by 1782 the wine from the area was being shipped to London. Wine merchants in France didn't like what they saw: competition. A group of them banded together and paid the Lords Proprietor, the overseers of the New Bordeaux project, £100,000 if support and subsidies for New Bordeaux would be stopped. The subsidies did stop, and so did commercial winemaking in South Carolina for eighty years. The death of Saint Pierre in 1784 also contributed to the end of New Bordeaux.

The next vineyards, planted by French in 1848, were to the south of New Bordeaux in Aiken. Truluck says in the 1860s the Benson and Merrier Winery in Aiken was famous for its claret, a dry red wine. Through the years other small wineries have come and gone, but nothing has come close to the brief glory South Carolina wines achieved in 1782.

Now, two hundred years later, in the land of soybean and cotton, dedicated grape growers are hard at work to add wine to the list of South Carolina's products.

Cruse Vineyards

Route 4, Box 404
Chester, SC 29706; (803) 377–3944

"We're in an unusual place. We were met with a great deal of skepticism," says Kenneth Cruse, owner of Cruse Vineyards. No wonder. Cruse owns the only operating winery in South Carolina. "I was the laughingstock of the community." He didn't have to put up with that for long, however. "When we harvested the fruit, the laughing stopped." And Kenneth says people were even more impressed once they tasted the wine.

The wine is very good, especially when you consider the grapes are coming from vines only planted in 1984. As vines grow older and mature, the fruit improves. Kenneth makes wines from both the European vinifera and from French-American hybrids. The wines include Seyval, Pinot Noir, Chardonnay, a blush, and a Proprietor Series of red, white, and blush. Wine sells for $6.75 to $14.

When Kenneth isn't making wines, he's working full time as the laboratory director at the local hospital. He believes it's the perfect background for a winemaker. "If you don't have an understanding of the basic microbiology, you don't know anything. I have been growing and fermenting things for a long time," says Kenneth. "I always wanted to do farming, but I realized it wasn't profitable [because the farmers] don't control the price. So I said, 'Let's plant a vineyard and eliminate all the people in-between.' You have to be a grape grower before you're a winemaker," says Kenneth. "The quality of your wine is directly related to the grapes."

Kenneth has been careful to begin slowly and not to get into debt. He built the house and devoted the basement to the winery and tasting room. When you drive up to the winery, it appears to be just your average residential home. But when you walk around back to the basement, instead of seeing a family's barbeque grill, you'll see winery equipment.

He planted the vineyard himself. "We're going slowly. We don't want to plant more than we can take care of." The winery is a three-person operation that includes his daughter, Lauren, and his wife, Susan, a viticulturist. With five acres now, they have no plans to expand in the near future. "We will let Lauren make that decision."

You can see the grapes on the Cruse tour. "I'll take visitors to the vineyard if they're interested. Then we allow them to taste whatever wine we're tasting that day." He'll treat you to a long or short tour and walk you around the basement, that is, the wine cellar. Kenneth or Susan will answer your questions. "When they come through the doors, they've come to see something," says Kenneth. "They've made a special trip, so I want to show them something."

DIRECTIONS: From Charlotte, drive south on Interstate 77 to exit 65. Drive west on Route 9 toward Chester for 5.6 miles. Turn right onto Cedarhurst for 1 mile. Turn right onto Wood Drive for .7 mile; winery is on your left.

HOURS: Tours and tastings on Friday from 3 P.M. until 6 P.M. and Saturday from noon to 6 P.M. Winery closed New Year's Day, Easter, Thanksgiving, and Christmas.

EXTRAS: Winery sells wine and wine accessories.

TENNESSEE

Highland Manor Winery

Beachaven Vineyards & Winery

Tennessee Valley Winery

Almost 100,000 gallons of wine were being produced in Tennessee in the late 1800s. Immigrants from Switzerland, Italy, and Germany brought their winemaking talents with them. They settled and made wine in the areas of Chattanooga, Memphis, Clarksville, Columbia, and Monteagle. Vineyards were scattered across the state. Wine production was still growing until laws were passed prohibiting the commercial sale of alcohol. The state went dry before Prohibition hit the rest of the country.

Even though families continued to make wine in their homes, commercial winemaking didn't return to Tennessee until 1980 when Highland Manor Winery opened. Legislation passed in 1977 had opened the door to small farm wineries in the state. The annual license fee was reduced along with the amount of tax wineries have to pay on each gallon produced. Since then other wineries have followed, with much success.

Grape growers harvest more than eighty types of the fruit in Tennessee. Some plant the native American vines, others use European vinifera varieties, and still others use French-American hybrids.

The biggest problems facing Tennessee vineyard owners are the cold winters and drastic temperature changes. When a thaw occurs sap starts flowing through the cane. Then if the temperature drops drastically, the vines can literally explode when the sap freezes. As vineyard owners reclaim the knowledge and experience of the growers of the 1800s, however, the wine industry in Tennessee will continue to grow and prosper.

Beachaven Vineyards & Winery

1100 Dunlop Lane
Clarksville, TN 37040; (615) 645–8867

"Champagne is what we were nationally known for as home winemakers," says Ed Cooke, who runs the Beachaven Winery with his wife, Louise. "We've been making wine for thirty years. We had this big hobby that was taking all my time." So in 1987 they opened Beachaven and turned the hobby into a business.

Judge William Beach, Ed's father-in-law, was instrumental in getting the winery started and worked there until his recent death. Ed says Judge Beach played a major role in rewriting the wine laws to make it possible for small wineries to operate in the state.

Although it's now a business instead of a hobby, Ed and Louise try not to be too businesslike, at least not with visitors. "It's a family operation," says Ed. "We try to give as much of a family touch as possible. That's the one thing we try to do is make people feel at home." Louise agrees, "After they sign in we tell them to look around and then have a tasting. We try to meet their needs."

Having a look around the Tudor-style winery means being able to see almost all the workings of the winery. Glass windows in the tasting room afford views of the press outside, a look into the lab, and a glimpse of tall stainless-steel fermentation tanks. "They can see what we're doing," says Ed.

If you want more than just a look, a tour guide will show you through the building. But watch out. You may end up working. "If people come in and are really interested, we'll put them on the bottling line," says Ed. "Their [spouses] can take pictures of them bottling. People love that."

If you're not in the mood to bottle, you can follow the stairs in the tasting room to the area where Beachaven makes its nationally known champagne. "This is a pretty nice place to send people down and explain the champagne and riddling process," says Ed.

Beachaven makes three sparkling wines in the traditional *méthode champenoise;* Brut, Sparkling Vin Rosé, and Sparkling Muscadine. All sell for $9.98. "We sell a lot of semidry wine," says Ed. One such wine is the Beachaven White, which Ed says is "a blend of a lot of different grapes."

From the sixteen-acre vineyard Beachaven produces such wines as the dry whites Riesling and Chardonnay, the dry reds Cabernet Sauvignon and Beachaven Red, and the dessert wines Blackberry and Muscadine. "Evidently we're the farthest north winery that sells Muscadine," says Ed of the native grape famous in the South. They also sell Gewürztraminer; "We tell everybody that's the only wine that goes with country ham and turkey," says Ed. The award-winning wines sell for $6.94 to $9.98.

As you leave the winery, stop and examine the outside of the building. The Cookes have added a most unusual feature to it—four inches of foam insulation. "So it's like a giant beer cooler," says Ed. But they've covered the "beer cooler" winery with tan and brown paint and turned it into a pleasant place to visit.

DIRECTIONS: From the Tennessee-Kentucky state line, drive south on Interstate 24 to exit 4. Turn east onto Highway 79. Turn right on Alfred Thun Road. Drive south for 2.2 miles. Turn right on Dunlop Lane and continue to winery. Driving north on Interstate 24 from Nashville, take exit 8. Drive east on Rossview for .6 mile. Turn left on Rollow Lane and drive 1.2 miles. Turn left on Dunlop Lane; after 1.5 miles winery is on the right.

HOURS: Tours and tastings available during the months of daylight saving time Monday to Saturday from 10 A.M. until 6 P.M. and Sundays from noon to 6 P.M. Winter hours are Tuesday through Saturday from 10 A.M. until 5 P.M. and Sundays from noon to 5 P.M. Winery closed New Year's Day, Easter, Thanksgiving, and Christmas.

EXTRAS: Winery sells wine and wine accessories. Picnic facilities.

Highland Manor Winery

Highway 127 South, Box 33B
Jamestown, TN 38556; (615) 879–9519

"This little room here is the champagne room," says Donna Wiggins, one of the tour guides at Highland Manor. And what a room it is. You'll feel as if you've been transported to a cellar in the champagne region of France. The old-world style room is bathed in a glow of yellow light radiating from lanterns attached to the red brick walls. Pillars topped with brick arches meet the low ceiling with its rough wooden beams. Stone floors support dark oak barrels. Bottles of fermenting champagne, tipped upside and resting on their necks, fill the room.

Only 600 to 700 bottles are made of the Champagne Royale. The winery has a list of people waiting to buy it and a list for the award-winning Muscadine wine. Although the winery continues to expand each year, owner and winemaker Irving Martin takes pride in the waiting list.

You will be able to taste and purchase the Chardonnay, Highland Red, White Riesling, Cayuga White, Royal White, Southern Blush, and Royal Rosé. The sweet style Royal Rosé is a blend of varieties such as Catawba and Alwood. The winery's driest wines are the Highland White, the Chardonnay, and the Highland Red. Most wines sell for approximately $6; the champagne costs $18.50.

"We grow 70 percent of our own grapes," says Donna. The winery owns twenty acres of vineyards, twelve acres more than it started with in 1977. "In the winter of 1984–1985 we lost eight acres of vines, so we replanted. It was about seventy degrees in December; then the temperature quickly fell to twenty-eight degrees below zero."

The winery offers a thirty-minute tour that includes an explanation of the winemaking process as you walk through the fermentation, aging, and bottling rooms. You'll see a piece of bark from the cork oak trees that grow in the Mediterranean region of Europe. "If you strip the bark off a tree around here you'd kill it," says Donna, explaining one of the differences between American oaks and cork oak trees.

Highland Manor, Tennessee's oldest operating winery, is nestled in green forests atop the Cumberland Plateau. You may buy the ingredients for a picnic at the winery, and the winery will provide a spot to enjoy it.

DIRECTIONS: From Interstate 40 at Crossville, take the Highway 127 exit (exit 317) and drive north for 28.7 miles. Winery is on the right.

HOURS: Tours and tastings offered year-round Monday through Saturday from 10 A.M. until 6 P.M., with tours ending at 5 P.M. During winter months winery closes at 5 P.M. Winery closed Sunday, New Year's Day, Easter, Thanksgiving, and Christmas.

EXTRAS: Winery sells wine and wine accessories.

Tennessee Valley Winery

15606 Hotchkiss Valley Road East
Loudon, TN 37774; (615) 986–5147

"People will soon hear of us," promises Chris Reed, a member of the Reed family that owns the Tennessee Valley Winery. Chris says that when people walk into the tasting room you can see by their faces that they think the wine will be terrible. But they are soon pleasantly surprised, and his brother, Tom, says, "We're definitely turning heads. We've got one of the best wineries in the Southeast."

The winery opened in 1984. "I needed a career, and my dad needed something for retirement," says Tom. His father, Jerry Reed, works as a pilot. "My brother and I do most of the work here. We're probably the youngest winemakers in the United States," says Tom. One brother is in his late twenties and the other in his early thirties. "We shock a lot of people. They're surprised that someone so young knows anything about wine."

It's probably not the age but the two men's attitudes that make people think they don't know anything about wine. They work hard at appearing to be just two "good ol' boys" from Tennessee who would be at home drinking beer and watching football. But they're not even from the state, and Tom has spent years learning about wine. "I went to SIU-C [Southern Illinois University-Carbondale], and worked on vineyards. I went to Mississippi State, and then I went to Tennessee. You have to have science back up your art, but it's basically an art."

They use their art to create Cabernet Sauvignon, Sauvignon Blanc, Classic Red Private Reserve, White Muscadine, Vidal Blanc, and numerous other wines. The wines sell for $6.49 for the White Riesling to $13.95 for the Cabernet. "We have fifteen or sixteen wines out there, and they'll find something to like," says Tom. And you'll be able to choose from a lot of award winners. "We were in the top 1 percent, that was with our Maréchal Foch," says Tom of a recent wine competition. "We just beat out some big vineyards. All of our wines have taken medals somewhere, Atlanta, New York, Dallas, Indianapolis."

The Reeds are glad success is finally coming their way. "It's taken a while to get the ball rolling," says Tom of the winery that produces approximately 12,000 gallons a year. "They basically told us we couldn't grow grapes in Tennessee." The Reeds have had a few problems with humidity and the diseases it brings but basically say they have had good luck with their thirty-acre vineyard.

The summer offers the best time to tour the winery. "Usually during harvest time we offer tours," says Chris Kimmins, who works at the winery. "That's when there's more to see, in July, August, and September. August is our biggest month." Any time of the year, if you catch one of the brothers at the winery, you can receive a full explanation of the winemaking process. You can also walk around the winery, look at the oak barrels used to age the wine, and see the 500-gallon stainless-steel tanks used for fermentation.

The Reeds didn't listen to what people said about growing grapes or attempting a winery in Tennessee, and now they have won more than 130 medals nationwide, including more than 50 gold medals. Not bad for a bunch of "good ol' boys."

DIRECTIONS: From Knoxville, take Interstate 40 west to Interstate 75. Head south to exit 76, Sugar Limb Road. Drive west for .1 mile and turn right. Winery is on the left.

HOURS: Tours and tastings offered all year Monday through Saturday from 10 A.M. until 6 P.M. and Sunday from 1 P.M. until 5 P.M. Winery closed New Year's Day, Easter, Thanksgiving, and Christmas.

EXTRAS: Winery sells wine and wine accessories.

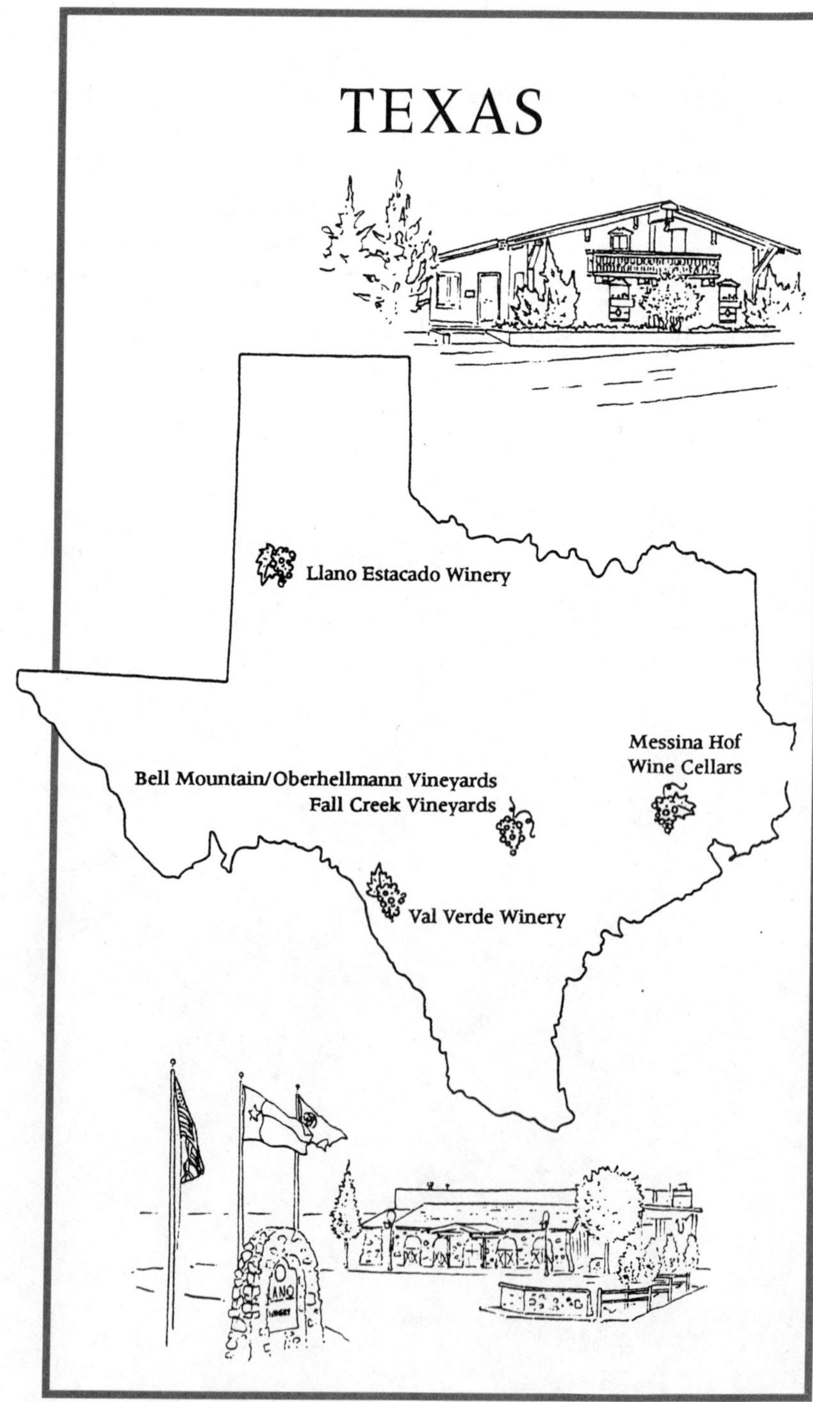
TEXAS
Llano Estacado Winery
Messina Hof
Wine Cellars
Bell Mountain/Oberhellmann Vineyards
Fall Creek Vineyards
Val Verde Winery

When most Americans think of Texas, they probably think of cowboys, big cars, J.R. Ewing, and oil—they certainly don't think of wine. But today's European wines owe their roots to Texas. Texan rootstock helped save the European wine industry when in the 1840s a fungus killed vines in England, Italy, and France. Europeans turned to the United States for rootstock that was phylloxera resistant and fungus-free. The first batch of rootstock sent to Europe was resistant to the fungus, but it carried the root louse phylloxera, which attacked other vines. All in all, more than 6 million acres of vines were lost.

Thomas Volney Munson, living in Denison, Texas, found native Texas rootstock resistant to louses. He sent the Texas plants to Europe. The rootstock was planted, and the grape-producing part of the plant was grafted onto it. Even today, the origins of some rootstock in Europe can be traced to a line of Munson's plants.

Two hundred years before Texas came to the rescue of European vines, the area now known as Texas had been producing fine wine. Franciscan friars brought their vines into the region from Mexico to make sacramental wine for their missions. As the Franciscans spread throughout Texas, so did vineyards.

Grapes grew well in Texas. In the early 1900s Texas wines were receiving prizes at international competitions. According to the Texas Department of Agriculture, twenty-six wineries were operating in Texas in 1900. But Prohibition closed all Texas wineries but one, Val Verde. Located in southwest Texas, Val Verde managed to stay open by selling grapes for home winemaking.

Texas now has wineries in every area of the state. More than twenty wineries helped produce a state total of almost a million gallons of wine in 1990, according to the Texas Department of Agriculture—this from a state that produced only 50,000 gallons eight years earlier.

With the help of irrigation, grapes have been doing well in vineyards with sunny days, cool nights, and sandy soil. The wines produced at these young wineries fare well in competitions with some of the best Californian wines. Quality and quantity will continue to increase if the Texas Grape Growers Association has anything to do with it. Its slogan is "Wine . . . The Next Big Thing From Texas."

Bell Mountain/ Oberhellmann Vineyards

HC 61, Box 22
Fredericksburg, TX 78624; (512) 685–3297

Although this winery is as well funded as the next, it hasn't yet gone "uptown." The tour reflects that. Tours are conducted by perky women who actually seem to be enjoying the opportunity to share their love for European-style wine. Even their costumes reflect that. They look as if they have just stepped off the set of a *Heidi* movie.

The tour begins on the steps of the winery with a brief overview of the history of the winery and winemaking. The guide tells you that she will walk you through the winery like a "bunch of grapes"—that is, just as the grapes go through the winery. As the tour proceeds the functions of each machine are explained. While all areas of winemaking are covered, no area is covered in depth. This moves the tour along at such a pace that even children won't have the opportunity to get bored.

The tour ends with a leisurely wine tasting. The tour guide remains with the group during the tasting, ready to answer any questions you may have.

The man behind the winery is Robert Oberhelman, a Dallas, Texas, food technologist, who planted his first vines in 1976. (He uses the German version of his name for the winery.) The trial vineyard was planted on the slopes of Bell Mountain with twenty-four varieties of grapes. Robert says, "Each year the ax fell on plants which could not measure up." He now has fifty-five acres planted with vines.

The area's sandlike soil and rolling hills, warm days and cool nights provide one of the best spots in Texas for producing wine. In 1986 the area was designated the Bell Mountain Viticultural Area, the first wine-growing appellation given in Texas.

From Bell Mountain grapes Oberhelman bottles a mix of red and white wines from the classic varieties such as Chardonnay, Sauvignon Blanc, Cabernet Sauvignon, and Pinot Noir. Most wines sell for less than $10.

DIRECTIONS: From Fredericksburg, drive north on Highway 16 for 14 miles. Watch for the winery's sign on the right.

HOURS: Open Saturdays from 10 A.M. until 5 p.m. from March 1 until the second week of December. Tours and tastings available Saturdays during those months from 10 A.M. until 4 P.M. every hour and half hour.

EXTRAS: Gift shop sells wine, wine accessories, and T-shirts.

Fall Creek Vineyards

Box 68
Tow, TX 78672; (512) 476-4477

You would probably expect to see stable doors in Texas, what with all those horses and cows around. But the doors you see at Fall Creek are not what you would expect—antique stable doors that once hung outside of Paris in the Louis Pasteur Laboratory.

You also may not expect to find so many award-winning wines. At the San Francisco Fair National Wine Competition in 1985, the winery won one of the first two awards ever accorded to a Texas wine. Since then its 1990 Chardonnay Texas Grand Cuvée won a gold at Les Amis du Vin International Wine Competition, its 1990 Sauvignon Blanc a gold at the Southwest Wine Competition, and almost all its other varieties have won medals in one competition or another.

You can imagine how proud that makes owners Ed and Susan Auler feel. The winery was founded in 1975 with a quarter-acre of vines, and to make their first wine in 1979, Ed and Susan had to borrow a winepress.

The Aulers began by planting Emerald Riesling, Chenin Blanc, Zinfandel, Carnelian, and Sauvignon Blanc. Later, they added

Cabernet Sauvignon, Chardonnay, Semillon, and Merlot to bring the vineyard up to sixty-five acres.

The Aulers don't rest on their laurels. Expansions have included a new visitors center and retail sales area. (Tow, Texas, was dry until a few years ago. The Aulers could make wine there but they couldn't sell it there.) The winery has storage capacity for 65,000 gallons, and 1990's production was 16,000 cases.

The winery is located on the ground floor of a 16,500-square-foot building. The second-floor living quarters, where the Aulers' friends stay when visiting, is decorated with antique French and English country furnishings. In 1985 author James Michener stayed in these rooms while researching his book *Texas.*

But the upstairs is not the only notable part of Fall Creek. During the winery's tour you can see the Pasteur stable doors, which lead to the room where the oak casks are stored. The thirty-minute tour begins in the visitors center. The group moves outside to the crush pad to hear about the history of the winery and the history of winemaking in Texas. Inside Fall Creek's winery you will hear how wine is made while you get a chance to look at the fermentation tanks. The tour continues through the bottling area, laboratory, and the oak room. After questions and answers you move back to the tasting room.

If the winery isn't too busy, the tour guide will stay with you during the tasting. If the guide needs to start another tour, the person pouring the wine will help you with your selections. Wines offered at the granite tasting bar include the regular Chardonnay, Sauvignon Blanc, Semillon, or Cabernet Sauvignon. Other Fall Creek wines are Emerald Riesling, Chenin Blanc, White Zinfandel, Granite Blush, or Granite Red. Wines sell for $5.99 to $13.99. Wines for sale and tastings depend on availability since many wines sell out.

Located in the beautiful hill country of Texas, this winery is a good choice if you're interested in visiting a high-class operation that consistently produces quality wine.

DIRECTIONS: From Austin, drive north on U.S. Highway 183. At Texas Highway 29 turn left (west) and continue until Highway 261. Turn right and then right again at Highway 2241. Fall Creek is 80 miles northwest of Austin, near Lake Buchanan. Winery is 2.2 miles northeast of the Tow Post Office.

HOURS: Tasting Room open Monday, Wednesday, and Friday from 11 A.M. to 2 P.M. Tours and tastings available Saturdays from noon till 5 P.M. Other times by appointment.

EXTRAS: Gift shop sells wine, wine accessories, and gifts.

Llano Estacado Winery

P.O. Box 3487
Farm Road 1585
Lubbock, TX 79452; (806) 745–2258

Llano Estacado must be the most talked about winery in Texas. It has been covered in such publications as *USA Today, Changing Times, The Wall Street Journal,* and *The Dallas Morning News;* and with good reason. In 1986 the winery's 1984 Chardonnay won a Double Gold from the prestigious San Francisco Fair. Some of the top award-winners in 1991 included the Signature White and Red, the 1989 Chenin Blanc, the 1989 Sauvignon Blanc, the 1988 Cabernet Sauvignon, and the 1989 Chardonnay.

Cellar master Mark Penna knows why the wines win so many awards: "Each wine gets special attention. We're not in a hurry to get them out. We use French oak. For our delicate white wine we prefer French. It's more subtle." It must be working, and not just because they win awards. "Real" people like the wine too. The wine sells across the United States and also in Europe.

The name of the winery has a 400-year-old history. Llano Estacado means "staked plains" in Spanish. When Spanish explorers traveled the land, they marked the vast plains with stakes as they sought the legendary Cities of Gold.

The winery opened in 1976. Although the winery has experimented with various types of grapes, it now uses only vinifera for winemaking. It planted its first commercial vinifera vineyards in 1978. In 1979 more vinifera was planted on the High Plains. Llano Estacado doesn't have too many problems growing vinifera in Texas. Although there is some danger of hailstorms and of late or early frosts, grapes ripen well, and in dry years there is irrigation.

In 1980 Llano Estacado bottled its first varietal wines under its label. A banner year for the winery was in 1990. That year the winery expanded its distribution in Europe and had its wine served at Camp David during a private luncheon with President Bush and then Soviet leader Mikhail Gorbachev. Its wine was also served at the International Economic Summit in Houston.

All this began from grapevines that were being used to provide shade for a patio. In the early sixties Texas Tech University had a small plot of experimental grapes. When the vines were pulled up to make room for construction, horticulture professor Bob Reed saved some of them and brought them home to plant near his patio cover, where the vines could grow and climb and provide shade. Without much care the vines flourished and produced a large quantity of grapes. Reed used the grapes for jellies and jams.

A friend of Reed's, C.M. McPherson, joined Reed and they began making wine. McPherson, a chemistry professor at Texas Tech, volunteered his basement as a temporary winery. They bought fifteen acres of land near Lubbock and began more experiments with different varieties of grapes. Then in 1975 other investors joined Reed and McPherson, and the Llano Estacado Winery was born.

The tasting includes three wines: a dry, a semidry, and a sweet. After tasting you may decide to buy one of the wines. Chardonnay sells for $13.50; Cabernet Sauvignon for $13.50; Chenin Blanc for $7.50; Merlot for $12; Llano Brut for $17; Riesling for $9; Late Harvest Riesling for $9; and Sauvignon Blanc for $8. Llano makes the blends Signature White and Signature Red, both for $8.50; Llano Red and Llano White, both for $6; and Llano Blush for $7.

You may enjoy a tour along with the tasting. Most tours run fifteen minutes, but if people ask questions, the tour may stretch to forty-five minutes.

Gather your questions and visit Llano Estacado. To learn how great wine is made, you might as well learn from the best.

DIRECTIONS: From U.S. Highway 87, drive 4 miles south of Loop 289. Turn left on Farm Road 1585 and drive 3 miles. Winery is on the right.

HOURS: Tours and tastings available Monday through Saturday from 10 A.M. until 4 P.M. and Sunday from noon until 4 P.M. Closed New Year's Day, Easter, Thanksgiving, and Christmas.

EXTRAS: The gift shop sells wine and wine accessories.

Messina Hof Wine Cellars

Route 7, Box 905
Bryan, TX 77802; (409) 778–WINE

Chenin Blanc and Miss Marple? If you are into the latest murder-mystery craze, where a murder is staged and you try to figure out who the "murderer" is, this winery is for you. Paul and Merrill Bonarrigo, owners of Messina Hof, will arrange a "murder" for you and your guests to solve. The characters in the murder mystery tie in with winemaking.

There are other special events held at the winery as well. Each fall it hosts Winefest. Visitors have the opportunity to stomp grapes as in the old days and to attend classes in cooking with wine. This is also when the Bonarrigos premier their new vintages. Write the winery for a complete listing of events, dates, and times.

When the Bonarrigos aren't managing murders, they are busy giving one of the best tours and tastings in the country. Paul and his staff treat the visitor as he does his vineyard—with pampering and tender loving care.

To begin the two-hour tour you are admitted into the Bonarrigo visitors center. The room has high ceilings with wide wooden beams. A massive fireplace takes up one side of the room.

When the visitors are gathered around the fireplace, the tour begins with a history of wine and how wine became intertwined with the Bonarrigo's family history. Paul Bonarrigo comes from a family with a traditional Italian love of wine. His grandfather owned his own vineyards in a small village outside of Messina, in Sicily. When the family came to America, they planted vines. Paul continues the tradition and shares it with you.

The tour moves out into the vineyard. For the next hour or so, the guide explains how the grapes are grown. Then it's into the winery's workrooms to follow the path the grapes take and to receive a description on how to make award-winning wine. Messina Hof wine has won more than 180 regional, national, and international awards. Some of the award winners include Johannisberg Riesling, Cabernet, Papa Paulo Porto, Chenin Blanc, and White Zinfandel. But the awards haven't gone to the prices; they are still a reasonable $5.49 to $14.99 per bottle.

Along the way the tour is filled with stories about wine customs; you learn, for instance, why the host always tastes the wine first. Years ago in England, serving poisoned wine was a common way to get rid of your enemies. Thus began the practice of letting the host drink first, so that the guest knew the wine was safe to drink.

The wine tasting can take up to an hour. But it is more than a tasting; it's like a wine seminar. After a wine tasting at Messina Hof, you'll be very knowledgable when selecting and tasting wine. The guide explains how to open the wine, how to order wine in restaurants, how to read a wine label, how to look for clarity in a wine, and how to taste-test a wine. At one point the whole tour group gargles wine! The visitors are told to let the wine sit on the top of their tongues. While the wine rests there, the group is instructed to cluck their tongues in order to release its full flavor.

The Bonarrigos began small; in 1977 they planted their first acre of vines. They now own thirty-three acres and manage eleven other Texas vineyards. The harvest produces about 50,000 gallons of wine.

The tour is informative and the Bonarrigos wonderful hosts,

but nothing can compare with the experience of being in a tour group learning to cluck.

DIRECTIONS: From Houston, take Highway 6 north. Turn right on Highway 21 and drive 2 miles east. Turn right onto Wallis Road and follow the signs to the winery.

HOURS: Winery open Monday through Friday from 8 A.M. until 4:30 P.M., with tours offered each day at 1 P.M. Weekend hours from 10 A.M. to 5 P.M. on Saturday and noon to 4 P.M. on Sunday. Weekend tours by reservation only; group tours by special arrangement.

EXTRAS: Gift shop offers a bridal registry and sells wine, wine accessories, deli foods, and books. Picnic facilities.

Val Verde Winery

100 Qualia Drive
Del Rio, TX 78840; (512) 775–9714

Not many cities can boast they have a winery right in the middle of them, but Del Rio can. But then the Val Verde vineyard has been there longer than the city. Located in southwest Texas, Val Verde is the oldest continuously operated winery in Texas. Frank Qualia founded the winery in 1883, and the Qualias have been running the winery ever since.

The Qualias even managed to keep the winery open during Prohibition. Using the railroads, Frank sold table grapes throughout the state. It is the only Texas winery that survived Prohibition.

The winery is still located in its original Spanish-style adobe building, with its twelve-inch-thick walls designed to keep the wine cool.

Even west Texas heat can't penetrate the adobe walls. As you walk into the winery the coolness surrounds you. With few windows to let in light, the darkness combines with the coolness for the feel of a true wine cellar. A few deep breaths of the musky smell of grapes, and the mood is set to do some serious wine tasting.

A fifteen-minute tour through the winery will provide a brief overview of the operation. On the tour you can see the 1880s Italian winepress that was the first one the Qualias ever used. After the tour there is a tasting.

Operated today by Frank Qualia's grandson Tommy, Val Verde is still a small, family operation. And Tommy plans to keep it that way. "My goal is to produce a great bottle of wine, not to sell wine in every store in Texas."

Helping in the wine production are Tommy's Toulouse weeder geese. These gray and white geese roam the vineyards eating the weeds that would take valuable nutrients from the vines. Although French in origin, the geese were brought to the winery thirty years ago from a farm in Missouri. Tommy says the geese are taken out of the vineyards "when their bills start turning purple." Listen for the honking and make sure you see the geese during your visit.

The geese wander through the 12 acres of grapes located behind the winery. There are also more acres planted in grapes thirty miles south of Del Rio. These acres have produced award-winning wines. A bottle of Don Luis Tawny Port, a port named after Tommy's father, is in the port museum in Venice.

Try the Don Luis Tawny Port for $9.95, and don't miss out on the other wines. For white wine lovers, the Chenin Blanc, priced at $6.95, is one of the best bargains in Texas. Other good deals include the Herbemont, the Chardonnay, and the Rosé of Cabernet, each for $6.75. For the price and the quality, when in Texas, do your wine shopping at Val Verde.

DIRECTIONS: From either Highway 90 or Highway 277, drive south on Main Street to Nicholson Street. Turn left and follow the signs to the winery.

HOURS: The winery is open Monday through Saturday. Tours and tastings available on demand Monday through Saturday from 9 A.M. to 5 P.M. Closed major holidays.

EXTRAS: The gift shop sells wine, wine accessories, gifts, and books. During August fresh grapes and grape juice are available.

VIRGINIA

Meredyth Vineyards
Oasis Vineyards

Prince Michel Vineyards

Barboursville Vineyards

Ingleside Plantation Vineyards

Mountain Cove Vineyards

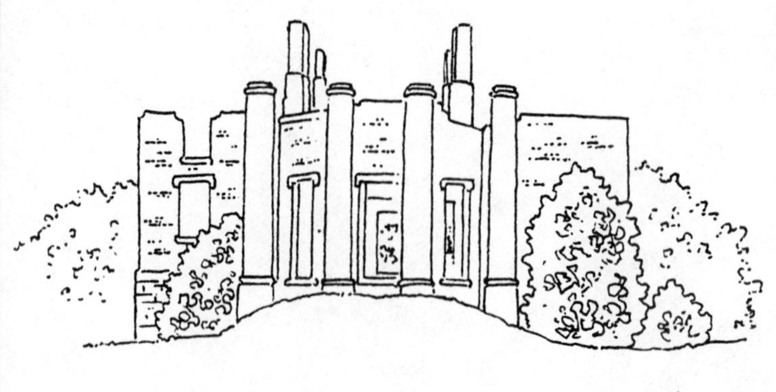

Enthusiastic doesn't begin to describe how Virginia winemakers feel about the future of wine in their state. You can feel the excitement at each winery. They're proud to be part of a tradition begun in 1607 when native grapes were first fermented by the Jamestown settlers. Lord Delaware, the governor of Virginia, is credited with making wine from European grapestock twelve years later.

Thomas Jefferson tried to grow French vines at Monticello in the late 1770s, but his attempts and those by others met with repeated failure as the vinifera vines didn't survive the new-world climate, soils, and diseases. People sought to find an alternative.

Success finally came from what most historians believe was an accident. Native grapes were inadvertently crossed with European varieties, and the Alexander, Catawba, and Norton were born. The Catawba went on to become the basis of the early Ohio wine industry, and the Norton became the popular grape in Virginia.

The vineyards fell into ruin during the Civil War and had no chance to recover before Prohibition shut down wine production. Only a handful of wineries produced wine in the state for the next fifty years. The passage of the state's Farm Winery Law in 1980 made it possible to produce wine without the previous heavy taxes and to sell wine at the winery. This opened the way for small wineries, and Virginians took advantage of it.

Hundreds of acres of vines may be found all over the state, and more wineries open each year. The winery owners' enthusiasm has been matched by the state's, and the state is supporting the wineries by placing signs with grape bunches on them on roads leading to the wineries to help visitors find them. October has been designated Virginia Wine Month, and most wineries participate by organizing festivals with grape stomps, music, food, and special tastings. It all provides a wonderful time for you.

Barboursville Vineyards

P.O. Box 136
Barboursville, VA 22923; (703) 832–3824

In the 1700s, when Thomas Jefferson tried to begin a wine industry in Virginia at his home in Monticello, he brought Filippo Mazzei from Italy to help him. Now 200 years later the Italians have returned to Virginia. The Zonin family, one of the largest wine producers in Italy, planted Barboursville Vineyards in 1976. They couldn't have picked a more beautiful site for the return.

Barboursville has 830 acres of rolling hills, 60 of which are in vines. According to tour guide and sales manager Christine Koontz, the rest support a cattle herd. The land was formerly known as Barboursville Plantations and was the home of James Barbour, governor of Virginia from 1812 to 1814. Barbour's mansion was designed by Jefferson and built between 1814 and 1822. Although gutted by a fire during the Christmas of 1884, most of the original brick walls still stand. You can get a glimpse of the hexagonal reception hall and the octagonal drawing room. Also still present are the unusual earthen ramps, which, instead of stairs, lead to the columned porticos at the front of the house. The ruins have been declared a Virginia historic landmark.

Barboursville Vineyards welcomes visitors around the ruins. You can sit back, sip some wine, and try to recreate from the ruins the grandeur of the mansion that once was the most highly appraised house in the area. Now the wine is some of the most praised in the state.

Barboursville offers whites such as Cabernet Blanc, Riesling, Chardonnay, Gewürztraminer, and Sauvignon Blanc; the reds Merlot, Pinot Noir, and Cabernet Sauvignon; and the Rosé Barboursville. The wines sell for $6.50 to $25.

During the tasting you may sample all the wines except for the Cabernet Sauvignon Reserve, which is produced in limited quantities. Tastings are offered Monday through Saturday, and thirty-minute tours are offered on Saturday. The tour begins with a brief history of the area while you're standing outside. Then it's inside the winery and a walk through the rooms that house fermentation tanks and oak barrels. The winemaking process is explained in great detail. "The types of oak as well as the size [of the barrel] are important in aging the wine. The winemaker uses Yugoslavian oak in search for a medium-body Cabernet," says Christine. As you walk past the oak barrels with the wine aging and mellowing in them, she points to them. "Those are his babies," says Christine of the winemaker who gives his wine loving attention. Jefferson would be proud.

DIRECTIONS: From Charlottesville take U.S. Highway 29 north to U.S. Highway 33. Drive east for 7 miles. Turn right on Route 20 for .2 mile, left onto Route 678 for .5 mile, and right on Route 777 for .3 mile. Winery is on the right.

HOURS: Tastings offered Monday through Saturday from 10 A.M. until 5 P.M. Tours available Saturday from 10 A.M. until 4 P.M. and weekdays by appointment. Winery closed Sunday, New Year's Day, Easter, Thanksgiving, and Christmas.

EXTRAS: Winery sells wine and wine accessories. Picnic facilities.

Ingleside Plantation Vineyards

P.O. Box 1083
Oak Grove, VA 22443; (800) 755–8294

Someone must have been looking out for Carl and his son Doug Flemer, Ingleside's chairman and vice-president. Just as they were looking for a winemaker, one sailed into their lives—literally. Jacques Recht, a French-trained Belgian enologist, and his wife, Liliane, were sailing around the world and were drawn to the Potomac River area after reading James Michener's *Chesapeake*. Upon meeting the Flemers, Jacques agreed to stay for three weeks to offer some advice and experience. Many years later he is still at Ingleside as the winemaker.

Vines were first planted in 1960, and the winery received its license in 1980. The vineyard has grown to fifty acres. The Flemers have had many years of agricultural experience. Ingleside is one of the largest growers of landscape plants in the East. The plants and vineyards grow on the 3,000-acre plantation, which is located between the Potomac and Rappahannock rivers.

There's much to do and see within 10 miles of the plantation. Five miles away is the George Washington Birthplace National

Monument. The center of attention there is a reconstruction of Washington's childhood home. Also at the monument are demonstrations of how people lived in the 1700s and their crafts. Eight miles from Ingleside is the Stratford Hall Plantation. The eighteenth-century home, stable, gristmill, and gardens fill the 1,600-acre estate once owned by the Lee family, two of whom signed the Declaration of Independence. The hall was also the birthplace of Confederate General Robert E. Lee.

Ingleside itself also has history. Constructed in 1832–1833, the original building has been a private academy, a garrison for Union troops during the Civil War, and a dairy. It is now a registered historic place. You can drive by the main house, which is .4 mile down the road past the winery entrance. Or you can stay at the winery and enjoy the eight-minute video about the plantation. "The viticultural region is Northern Neck George Washington Birthplace Viticultural Area," says the narrator. "And if you think that's a mouthful, wait till you taste our wine."

You may taste the wine before or after the video, and the winery also offers a twenty-minute tour. Ingleside produces more than ten types of red, white, rosé, and sparkling wine, selling for $4.99 for the George Washington Red to $9 for the Cabernet Sauvignon. There's also a Chesapeake Blanc, named for the area that brought winemaker Recht to Ingleside.

DIRECTIONS: From Interstate 95, take exit 45A, Route 3 east. Drive 34.2 miles east to Oak Grove. Turn right on Route 638; winery is 2.3 miles on the left.

HOURS: Tours and tastings offered Monday through Saturday from 10 A.M. until 5 P.M. and Sunday from noon to 5 P.M. Winery closed New Year's Day, Thanksgiving, and Christmas.

EXTRAS: Winery sells wine, and gift shop sells wine accessories, glassware, and local food products. Picnic facilities.

Meredyth Vineyards

P.O. Box 347
Middleburg, VA 22117; (703) 687–6277

"We're the oldest and most established," says Virginia Tucker Downs, the cellar master at Meredyth. "In 1972 we planted the first five acres. We have sixty in vines now." The vineyards take up a small part of the more than 200-acre farm that was once devoted to cattle and other crops. Archie Smith, Jr., owns the farm and Archie Smith, III, does the winemaking.

"We make eighteen different wines," says Virginia. "We have more wines than probably any other Virginia winery." The wines include Riesling, Chardonnay, Maréchal Foch, DeChaunac, and Blush. "This is the first Virginia Blush to win a gold," says Darlene Gaudreau, one of the tour guides. Most of the Meredyth wines have won awards, and they sell for less than $10.

Besides being proud of the wine, the personnel at Meredyth is also proud of the forty-five-minute tours. "It's a very informative tour," says Virginia. "We really take the time. It's a personal touch that we add." The tour takes you through the barn-turned-winery, from the lab where wines are checked for acid and sugar levels to the cellar where wines are cooled to stabilize them.

While many guides will tell you they age their wine in wood, the Meredyth guides will tell you why. "The French oak gives off a spicy, toasty flavor to wine. The American oak is buttery," says Darlene. And she'll tell you which wine is aged in which oak and why.

Back upstairs in the tasting room you may sample seven wines. Each wine is described, such as the Harvest Red. "It's called the red wine for the white wine drinker," says Darlene. During the tasting they provide cheese to munch between wines. You may also purchase the locally made cheese. "We sell other Virginia-made products," says Virginia. There's a full line of gourmet food, including wine jellies and herbal vinegars, for sale to take home or take on a picnic. And of course, there's wine for sale.

DIRECTIONS: From Washington, D.C., take Interstate 66 west to exit 8. Drive north on Route 245 for 1.1 miles. Turn right on Route 55 for 1 block. Turn left onto Route 626 for 3.8 miles. Turn right on Route 679 and drive 1.1 miles. (Route 628 branches to the right off of Route 679. Follow Route 679 to the left, and it becomes Route 628.) Follow Route 628 to the left; winery is on your left.

HOURS: Tastings available daily from 10 A.M. until 5 P.M. and tours from 10 A.M. until 4 P.M. Winery closed New Year's Day, Thanksgiving, and Christmas.

EXTRAS: Winery sells gourmet foods, wine, and wine accessories. Picnic facilities.

Mountain Cove Vineyards

Route 1, Box 139
Lovingston, VA 22949; (804) 263–5392

You're on top of the world, or so it seems, as you drive along the 105 miles of Skyline Drive perched on the Blue Ridge Mountains overlooking the Shenandoah Valley. The drive is dotted with hiking trails, picnic facilities, and scenic overlooks. The foliage in the fall draws crowds from all over the country, so that the drive sometimes suffers traffic jams. But that will give you the opportunity to drive slowly and enjoy the mountains, valleys, and deep ravines spread before you.

Beginning in Front Royal, the drive meanders from north to south before ending west of Charlottesville. "The Skyline Drive is just 15 miles from here," says Al Weed, winemaker at Mountain Cove Vineyards. The winery rests in the foothills of the Blue Ridge Mountains in an area Al describes as "a pretty location. It's one of the prettiest parts of Virginia. It's restful."

The winery's semisweet wines are named Skyline Red, Skyline White, and Skyline Rosé in honor of the area. Al also makes a dry white blend and a dry Harvest Red. The wines sell for $6. "My wines will tend to be slightly sweeter. I've really had a chance to see what people like," says Al. And most of the area residents prefer a semisweet over a dry. "I don't have any preconceived ideas of what wine should be," says Al. He believes that keeps him open to new ideas and to making wine for people, not for wine judges.

Most of that openness to new ideas comes from being a self-taught winemaker. Al didn't go to school to be told how to make wines. "I had a good job, making money, traveling. I wanted to farm. I thought I'd try it for five or six years, and here we are eighteen years later." Al has picked up a few things in those years. "I think I'm beginning to learn to grow things. I do all of the vineyard work myself."

The winery is operated by La Abra Farm, which is a corporation with nine stockholders. Al lives on the property and runs the winery. "I think the guys invested just so they'd have a place to go," says Al of the rural winery. The winery building, built into a small hill, is made with fieldstones. Inside it's all wood, from the oak paneling to the wood-slat floors.

Al planted vines in 1974. "We're one of the two oldest in existence," says Al. Only Meredyth Vineyards in Middleburg, established in 1972, is older. Working at one of the first wineries, Al has had to learn things by trial-and-error. "When we put vines in, in 1974 and 1975, we put in Cascade, and it was susceptible to a certain virus. We had to tear that out," says Al. "We've got seven varieties planted, all French-American hybrids."

Winemaking in the state is still young. "In Virginia we didn't have to just create wineries, we had to create vineyards. Basically we're selling a new product. There's things we won't know for twenty years," he says.

But Al does know how to make wine, and he will share his knowledge with you while he walks you around the winery on the forty-five-minute tour. It will last even longer if you'll take him away from bottling. "I like working in the vineyard. I don't like to bottle." But when he does bottle, Mountain Cove produces approximately 5,000 gallons a year.

When you visit the winery, gaze at the vista of the Blue Ridge

Mountains, listen to the birds, and *not* hear traffic, you'll understand what has kept Al here all these years.

DIRECTIONS: From Charlottesville, take Interstate 64 to exit 22A. Drive 28.2 miles south on U.S. Highway 29. Turn right on Route 718 and drive 1.6 miles. Turn right on Route 651; winery is on the right after 1.4 miles.

HOURS: Tours and tastings available daily April through December from 1 P.M. until 5 P.M. January through March the winery is closed on Mondays and Tuesdays but open all other days from 1 P.M. until 5 P.M. Other times by appointment. Winery closed New Year's Day, Easter, Thanksgiving, and Christmas.

EXTRAS: Winery sells wine and wine accessories. Picnic facilities.

Oasis Vineyards

P.O. Box 116
Hume, VA 22639; (703) 635–7627

"I call it an accident of faith. We fell in love with the land, and one thing just led to another," says Corinne Salahi of how she and her husband began Oasis Vineyards. "The only thing planned was that the soil was compatible. We have loamy soil, which is perfect."

Once the decision was made, the Salahis did plenty of planning. "We decided to go top of the line. We also started with hiring the top consultant. And I have a son at Davis now," says Corinne referring to the California university that offers one of the top programs in enology—the study of wine and winemaking—in the United States.

"It's a family operation," says Corinne of the winery. "I try to define that, and I think that means we're all stretched." Although the Salahis will be stretched a little further without their son, they won't let that get in the way of what they're trying to do at Oasis. "Our number one goal is to make the best wine with the fruit we have. We want to use all our own fruits. We don't want to make wine and sell it, we want to make wine with character," says Corinne. She adds that it's very hard to develop a character in wine that is made from grapes bought at different vineyards.

"Our vineyard is 13 years old. We have seventy acres of vines—fourteen varieties to pick," says Corinne. "Our production is about 30 percent sparkling." The winery produces approximately 15,000 cases a year, and in a completely underground building. "The walls and ceiling are 10 inches thick," says Corinne.

You'll be able to see the underground winery on the twenty-minute tour. "I go through the process. We cover the still wine and the sparkling wine. Sometimes I have people in a hurry so it's [the tour] five minutes." On the tour you'll see unusual fermentation tanks. "It's all concrete, 5½ inches thick," says Corinne of the round, squat tanks that are usually stainless steel. "They have a porcelain finish. They're common in Europe."

Upstairs in the tasting room a large stone fireplace makes an inviting area in the fall and winter. Little wood barrels with red pads on top provide seating. You can sit on the chairs by the fire, or if the weather's nice, you can sit outside and enjoy the Blue Ridge Mountains off in the distance.

"If they buy a glass of champagne, we will serve French bread and Virginia cheese," says Corinne. It sounds like a deal, with the champagne only $2.50 a glass. "We want people to come because of the stunning scenery. We have the most stirring view."

If you're not in the mood for sparkling wine, Oasis Vineyards sells other award-winning wines such as Gewürztraminer for $10, Chardonnay for $8, Riesling for $6.50, Cabernet Sauvignon for $12, blush for $6, and a rosé for $5. The champagne sells for $12.

Corinne says the vineyards make a perfect stop to a daylong or weekend trip to the area. "There's lots of antique shops and bed-and-breakfasts. We have the number one restaurant in the United States here. The Inn, in Washington, Virginia." Besides food, there's a feast for the eyes. The Skyline Drive rides the top of the Blue Ridge Mountains for 105 miles and overlooks the Shenandoah Valley. "The entrance to the drive is 5 miles away," says Corinne.

After visiting the sights of Washington and rushing around the city, the winery and western Virginia provide the perfect contrast. And they're only an hour away. It will feel like a world away.

DIRECTIONS: From Washington, D.C., drive west on Interstate 66 to the first Front Royal exit, number 28, Route 647. Drive south for 4.7 miles to Route 635. Turn right and drive 10.2 miles; winery is on the left.

HOURS: Tours and tastings daily from 10 A.M. until 4 P.M. Winery closed New Year's Day, Easter, Thanksgiving, and Christmas.

EXTRAS: Winery sells wine and wine accessories. Picnic facilities.

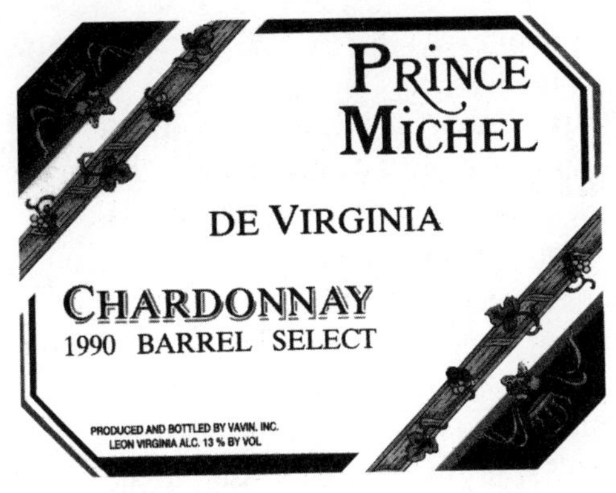

Prince Michel Vineyards

HRC 4, Box 77
Leon, VA 22725; (703) 547–3707

Prince Michel Vineyards receives its name not from one Michel but from several, and a few that were actual princes. One such Michel, Prince Michel, grandson of Michel II, was an emperor of the Byzantine Empire during the 800s. Michel III encouraged the planting of vineyards throughout Europe and enjoyed drinking wine.

The international name fits Prince Michel, Virginia's largest winery. Its principal owner is Frenchman Jean Leducq. Jean cofounded the winery with Virginian N.B. Martin and German Joachim Hollerith. Jean saw similarities with the French soil and climates and sought to make French-style wine in Virginia. He started in a big way. The 1986 production was 10,000 cases, and the winery has grown substanially since then.

Prince Michel makes two types of Chardonnay, both priced around $10. The Barrel Select ferments in French oak and the other Chardonnay is a blend, with two-thirds fermented in stainless steel and the other third fermented in barrels. Prince Michel's other wines include Cabernet Sauvignon; VaVin Nouveau; White Burgundy, a blend of Chardonnay and Pinot; and a Blush de Michel, a blend of Riesling and Cabernet. The White Burgundy and Blush de Michel sell for about $8.

Tastings are offered every half hour in the wine shop. While you wait you may watch a fifteen-minute video in the visitors center. A catwalk overlooking fermentation tanks and oak barrels provides the means to walk through the winery on the self-guided tour. Lighted displays explain each step of the winemaking process. You will want to spend some time looking over the pieces in the wine museum.

"It was my idea to have a museum make more familiar, to my American friends, the vine, the wine, and their art," says Jean on a sign in the wine museum, which is located near the entrance of the visitors center. One 6-foot-tall display case features tools and instruments used before machines did the work. The stake stirrup, for example, attached to the foot and was used to drive stakes into the ground. Old photographs, such as a late-1800s picture of grape harvesters in Napa Valley, line the walls. Best of all is the wooden winepress from Burgundy, France, dating back to the 1780s.

DIRECTIONS: From Culpeper, drive south on U.S. Highway 29 for 10 miles; winery is on the right. From Madison, drive north on U.S. Highway 29 for 8 miles; winery is on your left.

HOURS: Self-guided tours and tastings daily from 10 A.M. until 5 P.M. Winery closed New Year's Day, Thanksgiving, and Christmas.

EXTRAS: Gift shop sells wine, wine accessories, and international foods.

WASHINGTON

Mount Baker Vineyards

Arbor Crest
Wine Cellars

Château Ste. Michelle

Bainbridge Island Winery

Staton Hills Vineyard

Hinzerling Winery

Washington has two distinct grape-growing regions, separated by the Cascade Mountains. The Columbia River Basin, east of the mountains, has semiarid conditions. Most vineyards lie in this area. The west side of the mountains has cooler temperatures, and unlike the eastern side, enough rainfall to grow grapes without irrigation.

Washington's largest appellation, or federally designated growing region, is the Columbia Valley. The state's other appellations, Yakima Valley and Walla Walla Valley, are within the larger Columbia Valley appellation. All three areas are situated on the eastern side of the Cascades.

Washington has been growing grapes since the 1800s. Wine historian Leon Adams says that in 1872 Stretch Island in Puget Sound had grapes. And grape growing later became quite popular on Belle Island. Most plantings were in native American grapes such as the Concord. European vinifera was not planted successfully until the use of irrigation in the early 1900s.

Like most states, the boom did not hit until the 1960s and 1970s. Also, in 1967, André Tchelistchef met with the grape growers in Washington. A wine master and wine consultant to many of the industry's top wineries, he helped them discover what needed to be done for the state to reach its great potential.

Washington is now the second-highest producer of vinifera wines, after California. The state's vineyards have grown from a few hundred acres in the early part of the century to more than 11,000 acres. Vineyard owners will tell you it is because Washington is on the same latitude as northern Bordeaux and southern Burgundy. Temperatures are not the same, however, and Washington winters are much more severe.

Although winters are harder, Washington has been producing award-winning wines, garnering top honors over California and French wine. Winemaking is still relatively young in Washington, and winemakers predict an even greater future.

Arbor Crest Wine Cellars

North 4705 Fruithill Road
Spokane, WA 99207; (509) 927–9463

Not only will the wine knock your socks off, but so will the location. Arbor Crest's tasting room is located at the Cliff House, a national historical site. The house crowns a sheer 450-foot cliff that rises above the Spokane River. If the view of the Spokane River Valley doesn't satisfy you, the property will: one of the most bizarre estates around.

The carpet of green grass and brightly colored flowers is interrupted only by the unusual toys designed by the original builder, Royal Riblet. As you first enter the grounds, you see a test model of the Riblet Square Wheel. The wheel was made to fit any size or make of tractor. The inventor claimed it would work well in any terrain. The tread looks like that of an army tank. But its claim to fame is that with a square wheel, instead of a round one, more of the circumference is flat on the ground, giving better traction. Riblet believed it could be made for the same price as a rubber tire. Although the army initially expressed interest, it never got off the ground.

Your self-guided tour of the grounds takes you around a 16-

foot-square, red-and-black checkerboard. Wooden crowns are placed on the checkers when a player deserves a king. Riblet's grounds also include a swimming pool cut out of the basalt cliff, a lily pond, and a waterfall. Riblet's vista house, without sides and with a round roof for garden parties, affords a view of the Spokane River. The cement croquet court served two purposes. In the summer he added a thin layer of sand to keep the balls in place, and in the winter he flooded it with water for ice skating. There are also a putting green, an archery range, and a horseshoe-pitching area.

When Riblet and his friends tired of earthly fun, they took the tram for aerial fun. Built around 1927, the five-passenger tram went from his grounds, across the river, and down the 450-foot drop. Riblet would let his friends stop the tram in the middle of the ride and throw their fishing lines into the Spokane River. The tram was removed in 1956, but you may still see the terminal where the tram ride originated.

Riblet's house, now called the Cliff House, was built in 1924 in a Florentine style with a stucco exterior. The lower floor was made of the local basalt rock. Called one of the most electric houses of its time, Riblet installed an electric fireplace and built-in refrigerator. Don't miss the east side of the building where Riblet installed almost every shape and size of window. To make sure you catch all there is to see on the grounds, you can obtain a map of the area in the tasting room.

Born in Iowa in 1871, the inventor Riblet made his way to the Pacific Northwest to join the Riblet Tramway Company, owned by his brother, Byron. Royal also stayed busy inventing. He was said to hold approximately thirty patents. You can see some of the patent drawings in the winery's tasting room.

Arbor Crest bought the house and its seventy acres in January 1985, when much of the grounds had fallen into disrepair. Brothers David and Harold Mielke, owners of Arbor Crest, restored the house and use it for offices and a guest house. They also plan to build a new tasting room and a 100,000-gallon winery east of the vineyard. Because of the sheer drop, the property is closed to people under twenty-one years of age.

The people at Arbor Crest take pride in more than the grounds.

"We're proud of our whole line [of wines]," says Candace Frasher, who works in communications. "Our line tends to be high quality. Every single wine we make has the best quality. If the tank isn't what we want, we don't blend it. We just get rid of it. We are known to be the most award-winning winery, winning the most awards in Washington. We were named Washington state's 1988 Exporter of the Year." Arbor Crest sells wine to forty-three states and to such countries as Italy, France, New Zealand, Japan, and Korea—about a dozen in all.

Arbor Crest is also proud of what it will bring into the country. "We're the first U.S. company to import Russian wine. We'll import a red and a white," says Candace.

But of course the most pride is reserved for the wines of Arbor Crest, especially the Sauvignon Blanc. "This is our flagship at Arbor Crest," says Joe Algeo, head of national sales. Candace agrees, "It's put us on the map."

Arbor Crest's high international standing has also been increased with its award-winning Merlot and Select Late Harvest Johannisberg Riesling. The winery offers a full line of reds and whites, such as Chardonnay, Johannisberg Riesling, and Gewürztraminer, which sell for $5 to $11 per bottle. You may taste Arbor Crest's wine at the Cliff House.

Candace says the winery has had great luck with the Muscat Canelli, usually served as an after-dinner wine. "We serve [the Muscat Canelli] when we have special events," says Candace. She marinates fresh fruit in it. "Your guests will love you."

And you will love experiencing Arbor Crest.

DIRECTIONS: From Spokane, drive east on Interstate 90 to exit 287, Argonne North. Drive north on Argonne 1.7 miles to Upriver Drive. Turn east, right, and proceed 1 mile to Fruithill Road. Turn left; winery is .8 mile on the right.

HOURS: Self-guided tours and tastings from noon until 5 P.M. daily, but during the winter months, call first and check whether the road to the winery is passable. Closed New Year's Day and Christmas.

EXTRAS: Tasting room sells wine and wine accessories. Picnic facilities. No minors allowed.

Bainbridge Island Winery

682 Highway 305
Bainbridge Island, WA 98110; (206) 842–WINE

Getting to the Bainbridge Island Winery is half the fun. From Seattle, you take a 30-minute ferryboat ride to the city of Winslow, located on Bainbridge Island. Make sure you get out of your car and walk to the upper decks of the ferry. On a clear day, looking east, is Mount Rainier in its snow-capped splendor, and to the west are the cragged tops of the Olympic Mountains. And in the forward lounge, you can be entertained by musicians playing guitars and fiddles.

Then it's off to the winery in its neat blue barn with white trim. You can go immediately into the tasting room for some wine, or you can do the self-guided tour. The vineyard in front of the tasting room has five information panels at the row ends. Done in calligraphy, the signs provide better information than most guided tours.

For example, WHY WE GROW GRAPES HERE reads: "Wines from cool-climate regions, such as Loire of France or the Mosel of Germany, are world famous for their delicacy and fragrance." The temperature range at Bainbridge Island is similar to the Loire Valley, but with a drier growing season.

HOW WE GROW GRAPES HERE reads: "Our vines are not allowed to droop or hang as they are in hot areas where it is necessary to shade the grapes from sunburn. Instead, we train and trim our vines into upright hedges. This allows the sun to gently color the fruit and intensify the flavor."

The other signs bear such titles as WHERE WE GROW OUR GRAPES, GRAPES WE GROW, THE PERFECT CLIMATE, and WHY WE SELL OUR WINE ONLY AT THE WINERY. While you're reading the signs, keep an eye out for the three winery dogs and cat, Pinot, Chardonnay, Wombat, and Siegen.

WHY WE SELL OUR WINE ONLY AT THE WINERY, is the sign that greets you as you walk toward the tasting room, and it is the personal philosophy of the four people who work at the winery— JoAnn and Gerard Bentryn, the owners and their son Ian Bentryn, and Betsey Wittick. It says, "Most things we eat or drink are anonymous. We don't know where they're grown. We lack even the identity, never mind the personal philosophy, of the person who grew them. Wine from small wineries gives us a direct link with a specific place and a specific person: the vineyard and the winemaker."

If you're looking for a guided tour, the wine grower's tour is offered on Sundays at 2 P.M. On the hour-and-a-half tour, the winemaker will lead you through the vineyard and give a complete description from vineyard to bottling on how to make wine. Those interested may participate in pruning or harvesting, depending on the season. The winemaker braves rain or shine to give the tour every Sunday.

Take a minute as you walk toward the tasting room to notice the first cement step. Approximately 4 feet by 5 feet, the step has been beautifully decorated with grapevines, leaves, and grapes. As you enter the tasting room, you'll find more than just wine. Combining the personalities of an antique shop and wine museum, the tasting room houses a collection of old wine bottles, decanters, wineglasses, corkscrews, and books from France, Germany, and other countries. Glassware, with colors of every hue, fills the windows and cases. One vessel, a Persian amphora, is purported to date back to 2500 B.C.; an ancient Chinese *jue* (a bronze ceremonial wine-warming vessel) is from the Zhou Dynasty of the eleventh century B.C. Many of the items are for sale. Behind the tasting bar is a window where you can view the fermentation tanks and the goings-on of the winery.

When you visit "the only winery in the Puget Sound area that grows vinifera grapes," you'll be offered wines to taste, such as a Late Harvest Siegerrebe, Müller-Thurgau, or the Ferryboat White, a blend of Müller-Thurgau, Riesling, and Madeleine Sylvaner. The owners will explain the alcohol content, what percentage of sugar the wine has, and what style the wine was made in. They will also suggest food to accompany the wine. The wine, available only at the winery, sells for $6 to $20.

When asked why their winery is special, Gerard will tell you, "Locally grown, family owned. We like to think we provide the personal contact that you don't get with the larger wineries. You get to talk with someone involved with growing the wine."

Gerard, who spent two years in Germany in the military, recalls, "The milk was poor, and the wine was great." That's one way to get into the winery business. The Bentryns planted vines in 1978 and their vineyard has grown from 1.5 acres to about 7. Gerard doesn't see the winery getting too much larger.

"We don't want to get behind a desk," says Gerard. That's one of the reasons why he keeps production to around 5,000 gallons a year. Also it gives them the opportunity to try new things. "We're not limited in doing things that people aren't used to doing." Gerard produces some Müller-Thurgau in a dry style aged in oak as well as some of the traditional off-dry Müller.

"It's hard to get in the people's heads it's not where the wine is made or the wine produced, it is where the grapes are grown." The grapes the Bentryns use are grown in their vineyards on the island. You can tell they're proud of that and plan to keep it that way.

DIRECTIONS: From downtown Seattle at Pier 52, take the Winslow ferry to Bainbridge Island. Ferries leave approximately every hour all day long. Car and passenger for $6.65. Winery is located on the right, .25 mile north of the Winslow Ferry Terminal on Highway 305.

HOURS: Self-guided tours and tastings Wednesday through Sunday from noon until 5 P.M. Grower's tour offered on Sundays at 2 P.M. Closed New Year's Day, Thanksgiving, and Christmas.

EXTRAS: Winery sells wine and wine accessories.

Château Ste. Michelle

P.O. Box 1976
14111 NE 145th
Woodinville, WA 98072; (206) 488–1133

"California can take care of quantity, and we'll take care of quality," says a guide at Ste. Michelle. The winery may concentrate on quality, but it hasn't done too badly with quantity either. Château Ste. Michelle produces millions of gallons of wine.

In 1934, just after the repeal of Prohibition, the winery was bonded in Grandview, Washington, where it concentrated on sweet dessert wines made from fruits, berries, and grapes. It has come a long way since the 1930s. Ste. Michelle planted vinifera vines in 1950, and its current 2,700 acres of vineyards, located in eastern

Washington, reflect its focus on Cabernet Sauvignon, Chardonnay, Merlot, and Riesling. In 1967 it began producing exclusively the classic vinifera wines and took its present name.

It was quite a milestone for Washington wineries when Château Ste. Michelle's 1972 Johannisberg Riesling was judged number one when competing against Californian and German wines.

You can visit Ste. Michelle's wineries at Woodinville or Grandview. The Woodinville Winery is the corporate headquarters of Stimson Lane Wine & Spirits, Ltd. at the old Hollywood Farm. Its eighty-seven-acre grounds are filled with trees, flowers, fish ponds, and vineyards. The landscaping, designed by the Olmsted family, who also designed New York's Central Park, has been replanted and groomed to its original glory.

When you turn into the winery, you'll drive past vineyards. "Tradition says you enter a winery through its vineyards—so it's tradition," says a guide. Past the vineyards and down the tree-lined walkway is the winery, wine shop, and reception area, where the tours begin on the half hour. While you are waiting there's plenty to do. Panels on the wall of the lobby are filled with information on wine history and on vineyards, and plastic containers show different types of soil found in the vineyards.

The forty-five-minute tour takes you through the lobby and into the back of the winery where fermentation takes place. The group makes several stops as the guide discusses fermentation, crush, and bottling. The barrel room is quite impressive. Ste. Michelle uses four French and two American types of wood barrels. After the tour it's off to the tasting room. During the tasting your guide will talk about food that accentuates the wine.

The Fumé Blanc is made in the style of dry whites from the Loire Valley of France and is fermented and aged in the French oak. The insides of the barrels have been charred, and this lends a smoky flavor to the wine—thus the French word *fumé,* for *smoke.* Or try the famous Johannisberg Riesling, the wine that brought the winery much recognition. Other Ste. Michelle gold-medal winners include Cabernet Sauvignon, Chardonnay, Chenin Blanc, Gewürztraminer, Merlot, Muscat Canelli, Semillon, and Late Harvest White Riesling.

Perhaps it's because the winery is so popular, or because it sells wine in almost every state and sales at the winery are not important, but whatever the reason, the tasting is rushed. Don't expect to spend much time exploring the complexities of the wine. Before the pourers have given everyone a glass and you have had a chance to taste the wine, they are talking about the next one. You'll probably prefer buying a bottle and leisurely discovering its flavors under one of the shade trees or next to the fish pond.

DIRECTIONS: From Seattle, take Highway 405 north to exit 23 east, which is Highway 522. Continue east on 522 for .8 mile to the Woodinville exit. Turn right and proceed to NE 175th Street. Turn right onto 175th and continue to Highway 202. Turn left and drive 1.7 miles. The winery is on the right.

HOURS: Tours and tastings available daily from 10 A.M. until 4:30 P.M. Closed New Year's Day, Easter, Thanksgiving, and Christmas.

EXTRAS: Winery sells wine. Picnic facilities.

Hinzerling Winery

1520 Sheridan Avenue
Prosser, WA 99350; (509) 786–2163

"We have the '12-360' tour. It takes twelve seconds. You stand and turn around 360 degrees, and you've seen it," says owner Mike Wallace of the small winery. You won't be led around huge tanks, but while you taste the wine, surrounded by winery equipment, Mike will tell you how and why he makes the wine.

What the winery lacks in a long walking tour, it makes up for in history. Mike produced his first wine in 1976, making Hinzerling the oldest continuous family-owned winery in the Yakima Valley, and the third oldest in the state. "Twenty some odd years ago I was in the service in California," says Mike about when the wine bug bit him. He read that research was going on in Washington about grape growing, and he thought the state was right for him. "In 1971 I bought land, and in 1972 planted vines. I started a vineyard because there wasn't any grapes."

After years of making wine and winning gold medals, Mike was ready for a change in late 1987 and early 1988. His parents, who helped him at the winery, wanted to retire, and the winery was too much for Mike alone. He sold the winery and vineyard and spent his time as a wine consultant. When the buyer had financial and other problems, Mike took the vineyard and winery back. Then wanting to spend more time making wine, Mike sold the vineyard. "I'm experimenting. I specialize in small lots of things—ports, dessert wines, reds like Cabernet and Merlot."

Mike warns visitors that if they like bland watered-down wine, this isn't the place to visit. "We're not going to be something everyone will like. We make wine with lots of flavor. A little bit out of the mainstream. Something you'd find interesting. It's an interesting little niche."

Hinzerling's whites are Ashfall White for $5.49 and Dry Gewürztraminer for $5.99. The reds are 1989 Pinot Noir for $11.99 and Cabernet Sauvignon vintages from 1979, 1982, 1983, 1984, 1987, for $10 to $24. His dessert wines are Collage, Rainy Day Fine Tawny Port, Three Muses Ruby Port, and 1989 Vintage Port. Prices are $11.99 to $15.99.

Mike's parents couldn't stay away either. Either Mike or his parents will aid you in a wine tasting. "Someone directly involved in winemaking [gives the tasting]. People taste pretty much what they want to taste. We'll do a barrel tasting [take the wine out of the barrel it's aging in] if they show some interest," says Mike.

Although the winery is closed from Christmas to April, giving Mike a chance to catch up on his work, if you're in the area and want to try some wine with a lot of flavor, try knocking on his door. If he's in the winery working, he'll offer you a tasting. Mike says, "Stop by and take a chance."

DIRECTIONS: From Yakima, drive southeast on Interstate 82. Take exit 82. Turn left and drive 1.2 miles on Wine Country Road. Winery is on the left.

HOURS: Tours and tastings daily from April through December 24, 11 A.M. until 4 P.M. Closed Thanksgiving and from Christmas until March 31.

EXTRAS: Wine and wine-related gifts sold at winery.

Mount Baker Vineyards

4298 Mount Baker Highway
Everson, WA 98247; (206) 592–2300

You can visit the wineries in eastern Washington with their dry, hot, desertlike conditions, or you can visit Mount Baker Vineyards in western Washington and sit among the green, vine-covered hills under the shadow of snow-covered Mount Baker and the Twin Sisters Mountains. And you can enjoy a great production tour.

The tour begins where it should, in the vineyard, where fifteen varieties of grapes are grown. "When they bloom, the fragrance is incredible," says Al Stratton, winery owner. "It's almost like roses. Delicate and sweet. The tiniest little blossoms. About the size of a pinhead."

Al talks about the latitude and the warm air the valley receives. "It will be cloudy except by the vineyard. The Nooksack Indians called it 'hole in the smoke,' for this small area of the valley," says Al.

The valley is the Nooksack Valley, an area where most people believed wine grapes couldn't be grown. But Al and his wife, Marge, thought otherwise. They have found that the Nooksack Valley provides a mild, 240-day growing season. Classic German varieties, along with Chardonnay and Pinot Noir, have done well in this area. The Strattons, working with Washington State University, have planted almost a hundred varieties in order to find the best for the region. Twenty-five acres of vines were first planted in 1977. Production has increased steadily.

From the vineyards you move into the back of the winery to hear about crushing and fermentation. "Young fermenting wine is very unique," says Al. "It takes anywhere from five to ten days to ferment."

Then, it's into the tasting room for more fun. A board behind the tasting bar has the wines listed, and the wines with a gold seal by them are the ones open for tasting.

The winery produces a Müller-Thurgau in a dry style. "We made it especially to go with the fish and salmon from the area. Your grilled fishes or smoked foods. We lightly oaked it," says Al. Then there's Madeline Angevine for $6. "This is for your more delicately flavored fish or poultry." Or the Riesling, "It's great with oysters on the half shell."

Mount Baker also sells Siegerrebe, Gewürztraminer, Okanogan Riesling, Chardonnay, Cabernet Sauvignon, Pinot Noir, plum wine, and select white blends. The excellent Müller-Thurgau is a great buy at $6.

This winery knows how to take care of everybody. Those who don't drink may try their Old World Nectar grape juice made from selected wine grape varieties.

DIRECTIONS: From Interstate 5, take exit number 255 in Bellingham. Drive east 11.5 miles on Highway 542 to Hillard Road. Turn left onto Hillard; winery is on the left.

HOURS: Tastings and tours available Wednesday through Sunday from 11 A.M. until 5 P.M. Closed New Year's Day, Easter, Thanksgiving, and Christmas.

EXTRAS: Tasting room sells wine, wine accessories, locally made ceramic bottles, jams, berry syrups, and vinegar. Picnic facilities.

Staton Hills Vineyard

71 Gangl Road
Wapato, WA 98951; (509) 877–2112

As you enter Staton Hills Vineyard, you pass through rows and rows of grape vines. You'll notice types of trellises not seen at other wineries. Take time to look at the different trellis systems being used. The trellis system, costing three times that of conventional systems, has approximately 144 miles of supporting wires, and the wires have from 210 to 240 pounds of tension. David and Susanne Staton use these trellises to expose more new grape leaves to the sun for better growth for the vines and a higher quality grape.

Past the trellises you'll see the beautifully landscaped grounds, an oasis in the brown, dry Yakima Valley. Drink your wine next to the fountain or bring food to cook in the barbecue pits the Statons provide on the grounds.

The tasting room, at the front of the winery, is a beautiful building with a native-stone bottom half and a cedarwood top. Large, double wooden doors decorated with grape bunches and leaves open onto the impressive tasting room.

The first floor has a massive two-story stone fireplace, wood beams on the ceiling, and a beveled cathedral wood ceiling. An unusual addition to the tasting room is a 12-foot-tall grape bunch made from purple balloons. "We just keep adding [balloons] and then they shrink up, just like little grapes do," says Debbie Arcand, banquet manager and part-time wine pourer. Sure enough, the older, shriveling balloons add a flavor of authenticity, looking like grapes left on the vine too long. On the floor are piles of woven baskets of every size, shape, color, and color combination. "Susanne Staton had a friend go to China, and she shipped them back," says Debbie.

A loft on the second floor is used for a banquet room. It also provides a great view of the Yakima Valley and Mount Adams. The other wall of the banquet room has a glass window from which you can see fermentation tanks, oak casks, and the catwalk that runs through the winery. The winery produces more than 100,000 gallons of wine, including a sparkling wine for $15.95, a Late Harvest Riesling for $9.50, and a Chardonnay for $12.95.

You can tour the winery by appointment. Or you can just stop by, and while you're tasting the four complimentary wines, the pourer will give you a taste tour, explaining the wines you taste. For instance, you will learn that many of their white wines are aged in oak.

The whites may be good, but the winery is betting its future on the reds. "One thing we've found out about this site is that it makes a great Cabernet Sauvignon," says winemaker Rob Stuart. And others agree. Staton Hills Cabernet has been a consistent winner in the American Wine Competition. The Cabernet Sauvignon and the Pinot Noir have won numerous gold medals. Rob says, "I think Washington state has the ability to grow world-class wine."

DIRECTIONS: From Yakima, drive 4 miles south on Interstate 84. Take the number 40 exit. The winery is right off the highway on the northeast corner.

HOURS: Tastings Tuesday through Sunday, 11 A.M. until 5:30 P.M. during the summer months and noon till 5 P.M. during the winter months. Tours by appointment only. Closed New Year's Day, Easter, Thanksgiving, and Christmas.

EXTRAS: Tasting room sells wine, wine accessories, food for picnics. Picnic facilities.

WEST VIRGINIA

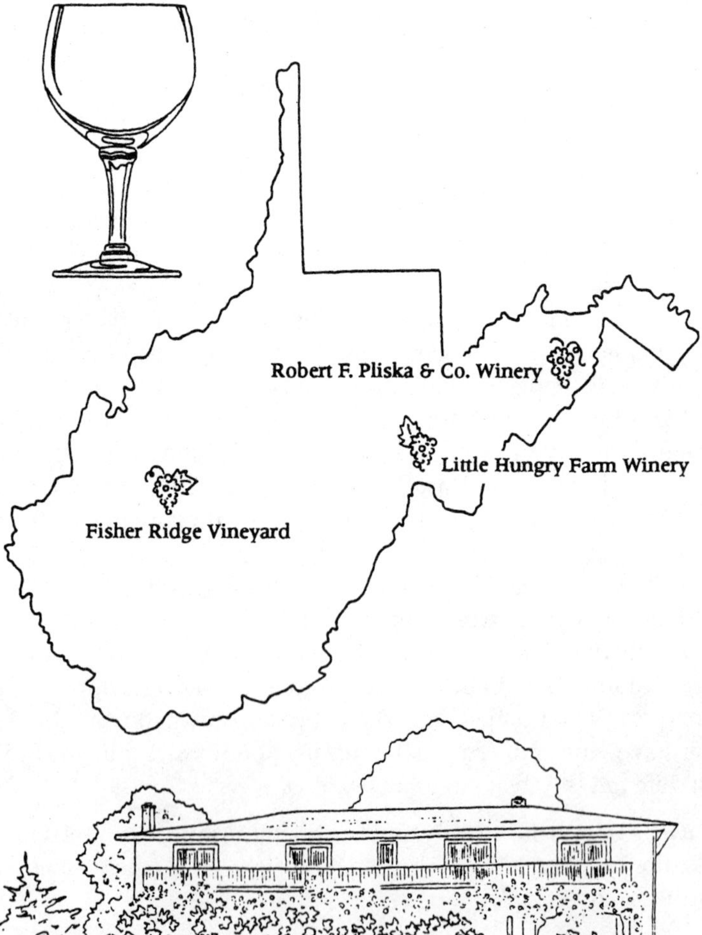

Robert F. Pliska & Co. Winery

Little Hungry Farm Winery

Fisher Ridge Vineyard

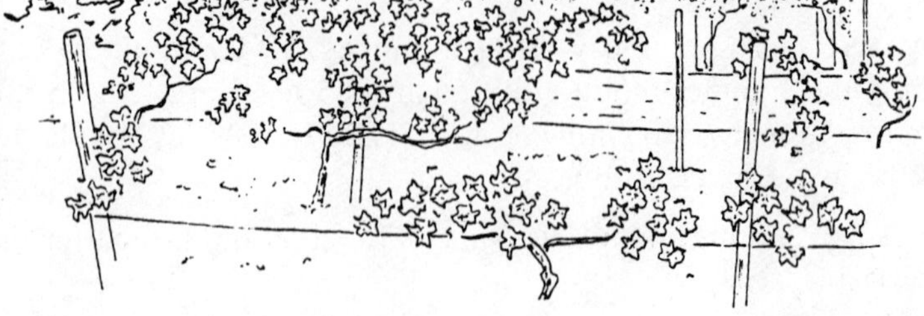

West Virginia has approximately five species of native wild grapes and has a long history of using them for table grapes and jellies. Records of winemaking date back to the early 1800s and show wine being produced in the Charleston area around 1825.

The most well-known winery of the 1800s was Thomas Friend and the Friend Brothers Winery, which was operating commercially in 1856. The wine cellars, dug into the earth like caves, are located in Dunbar, a town just west of Charleston. The cellars have been restored, and a park has been built around them. It's an ideal site to visit if you're interested in the history of winemaking in the United States.

The Civil War interrupted local wine production; vineyards either became battlefields or fell into disrepair as vineyard workers went off to war. Even so, wine was being produced in more than ten West Virginia counties as late as 1880.

Wine production only recovered from Prohibition and the Prohibition era in the 1970s and 1980s. With the passage of the farm winery law in 1981, wineries may now serve samples and sell wine at their locations. Grape growing, too, has recovered, and now more than half of the state's counties grow grapes—either for table use, jams and jellies, or for wine production. Each year more wineries open as people become more interested in local products, try the state's wine, and support existing wineries. Grape growers and winemakers are optimistic that West Virginia will soon be famous for wine.

Fisher Ridge Vineyard

Fisher Ridge Road
Liberty, WV 25124; (304) 342–8702

"We were the first in one hundred years," says Wilson E. Ward, owner of Fisher Ridge, the first winery licensed in West Virginia since Prohibition. "I started as an amateur. We started with about 500 gallons. Now we're a 5,000-gallon winery."

Although he was the first, Wilson doesn't want to be the largest. "My thrust isn't to get very big. Essentially what I'm trying to do is [focus on] 'How good can the product be?' Our aim is to establish credibility of growing quality wines. I think we're doing all right." Wilson believes that as the state grows as a wine-producing area, its wines will improve with age. "We get better. The vines improve. Your skills as a winemaker improve."

The skills must be improving. Fisher Ridge's wines won seven medals in the 1992 annual West Virginia wine competition. Wilson's Cellar Master Red, Vidal Blanc, and Maiden's Kiss all won gold medals. The Cellar Master Red also won best of show in its category.

But wine growing in West Virginia is not without its problems.

"We have all sorts of time to grow, and plenty of heat to grow, but we have low winter temperatures," says Wilson. Some vines crack and split open when temperatures go below freezing.

And that's not all. "Your biggest enemy is the fungus," says Wilson. Humid summers promote the growth of fungus, which attacks and kills the plants. To fight fungus, Wilson places the vines on high trellises so breezes may keep the humidity—and hence fungus—away from the vines. "We planted in 1977. The vines are trained up real high. The recommended height is six foot, but I'm only five foot seven so I put them on my eye level," says Wilson. "Wind movement keeps things dry."

Wilson will be happy to show you how he does it; just call for an appointment. "Best days are Wednesdays and weekends," says Wilson. His duties as a full-time dentist keep him from offering tours daily. But when he does give a tour, he's pleased to show off "all the things that Gallo has, except on a smaller scale."

Wilson produces some wonderful wine, such as his best sellers, a blush blended from Cabernet Sauvignon and Seyval and a Blanc de Blanc blended from Chardonnay, Seyval, and Vidal. He also makes Chardonnay, Vidal Blanc, and a German-style wine called Maiden's Kiss. Wines sell from $5.25 to $6.95.

DIRECTIONS: From Charleston, drive north on Interstate 77 to exit 124, Kenna. Turn left onto Route 34 South and drive for 8.7 miles. Turn right on Fisher Ridge and drive 2.6 miles. Winery is on the left.

HOURS: Tours and tastings by appointment.

EXTRAS: Winery sells wine and wine accessories.

Little Hungary Farm Winery

Route 6, Box 323
Buckhannon, WV 26201; (304) 472–6634

"Everything that's in good wine plus three quarters of a pound of honey per gallon," says Ferenc "Frank" Androczi about the wine he makes called Melomel. "They call it honey mead if you make it with water and honey." But Frank uses wine to ferment along with the honey, so it's ready to drink sooner than if you used water. "They call it ambrosia. I cure myself with just this," says Frank in his broken English.

He swears it's the "Fountain of Youth, Health, Strength, and Beauty." He tried to use that on his label, but the federal authorities wouldn't allow it. Frank has plenty of stories he'll tell you about run-ins with the government. Or he'll just tell you stories:

"In the ancient world, when formal marriage was unknown, the boy grabbed the beautiful girl and hid her in the woods. The boy's parents and brothers knew where they were, however, and during the night they took them bread, honey, and honey mead for their well-being and support. When the girl's parents and brothers stopped looking for her, or when they didn't want her back, then the young couple returned to the tribe.

"Usually the stay was for one month and the Greeks called it 'Moon'; from here came the word used today, *honeymoon*. If the bride happened to come back pregnant, the credit was given to the mead maker and not to the husband, since honey was the symbol of fertility."

Frank says he's never tasted aspirin. He believes he's never needed it because he drinks Melomel. And he says if you drink it, "you'll live to be 150 years old." Frank is a great testimony to his product. In his mid-seventies, you'll have a hard time keeping up with him as he bounds through the vineyards and orchards explaining which plant is which and how he cares for them.

Born in Hungary, Frank immigrated to the United States in the fifties. He came to West Virginia in 1967 to teach library science at Wesleyan College. Frank thought the state was perfect for grape growing, so he planted a vineyard. After retiring in 1981 he turned his full attention to producing Melomel.

Frank has vineyards around his house in Buckhannon as well as a larger winery and three acres of vines and fruit trees 7 miles northeast of town in Kesling's Mill. Frank will gladly drive to the winery and show you his operation there. He'll also offer more samples of his wine, which sells for $6 a bottle. Frank only sells Melomel, but he makes several types, some sweeter than others. He uses his own honey to mix with his wine to make Melomel. "I have bees in four locations," says Frank.

He uses a secret recipe he learned from his family, although if you ask him he'll tell you he's made some improvements. "My mother said to my father, talking about my wine, 'George, this is very good.' He didn't talk to me for three days. He was very proud. He said he made the best in the world, but he was wrong; he never tasted mine."

Frank will keep you busy for hours with his stories and wine samples. Women are warned that he has been known to steal a kiss. If nothing else, he'll steal your heart.

DIRECTIONS: From Interstate 79, drive east 12.3 miles to Route 20. Turn right and drive for 1.8 miles. Winery is on the left, behind Good Deal Used Autos and across the street from Krogers. The entrance is hard to see.

HOURS: Tours and tastings anytime, but call first to make sure Frank is home.

EXTRAS: Winery sells wine.

Robert F. Pliska & Company Winery

101 Piterra Place
Purgitsville, WV 26852; (304) 289–3900 or (304) 289–3493

"God created the grape to be wine," says Bob Pliska, winemaker at Pliska and Company Winery. "Wine starts with the grapes. If you keep the heat away, keep it clean. You try to channel the grapes along its best route. I'm a wine keeper. That's basically what winemakers do."

If it sounds to you as though Bob is a philosopher, you're right. And the winery reflects that. Bob and his wife, Ruth, named their vineyard, planted in 1974, *Piterra.* The word comes from the Greek letter *pi,* which they say means "never-ending," and the Latin word *terra,* which means "land." Gazing over the rolling vineyards nestled in the foothills of the Appalachian and Allegheny mountains, you come to appreciate the meaning of "never-ending land."

Bob and Ruth have a deep concern for the mentally handicapped. Inspired by their son, who has Down's syndrome, they built a home for special-need adults. Called The Vineyard, it is located near the winery and receives some funding from the wine sold there.

"We put on a wine auction for the home for the handicapped in Chicago and Washington, D.C. They do a lot to try and keep the home going," says Bob. He sees a direct connection between West Virginia wineries and special-need adults. "They have a lot in common. Both are trying to establish their self-worth."

The winery traces its origins to when Bob and Ruth moved to the region less than twenty years ago. "I was in the Navy, and we were looking for a place to retire," Bob says. Several tours to Washington, D.C., gave him the opportunity to discover Purgitsville. "We recognized we couldn't grow corn here, and apple orchards were going bankrupt. We thought there had to be something to do with the land," says Bob. And from the land and the vineyard Bob came up with the wines Aurora, Seyval, Chancellor, Foch, Concord, and the blends August Gold, Ridge Runner Red, Ridge Runner Gold, and Rosé. Mountain Mama is the apple wine. The wines sell for $3.95 to $8.93.

Your interest will determine the length of the tour. "Some people just want to taste the wine. It's different [the tour] for different people," says Bob. There's a ten-minute video. Then Bob will show you around the winery and pour some of the wine that he's proud of. "Our Foch was used in Paris to celebrate the 200-year anniversary of Thomas Jefferson's visit," he says. It appears that people are giving Bob's wine respect while he's busy helping other adults gain theirs.

DIRECTIONS: From Cumberland, Maryland, on Interstate 70, take the U.S. Highway 220 exit and drive south for 39.8 miles. The winery is on your left.

HOURS: Tours and tastings available from April 15 to October 15, Tuesday through Saturday from 1 P.M. until 4 P.M. The rest of the year by appointment only. Winery closed from October 16 to April 14 and Easter.

EXTRAS: Winery sells wine and wine accessories. Picnic facilities.

WISCONSIN

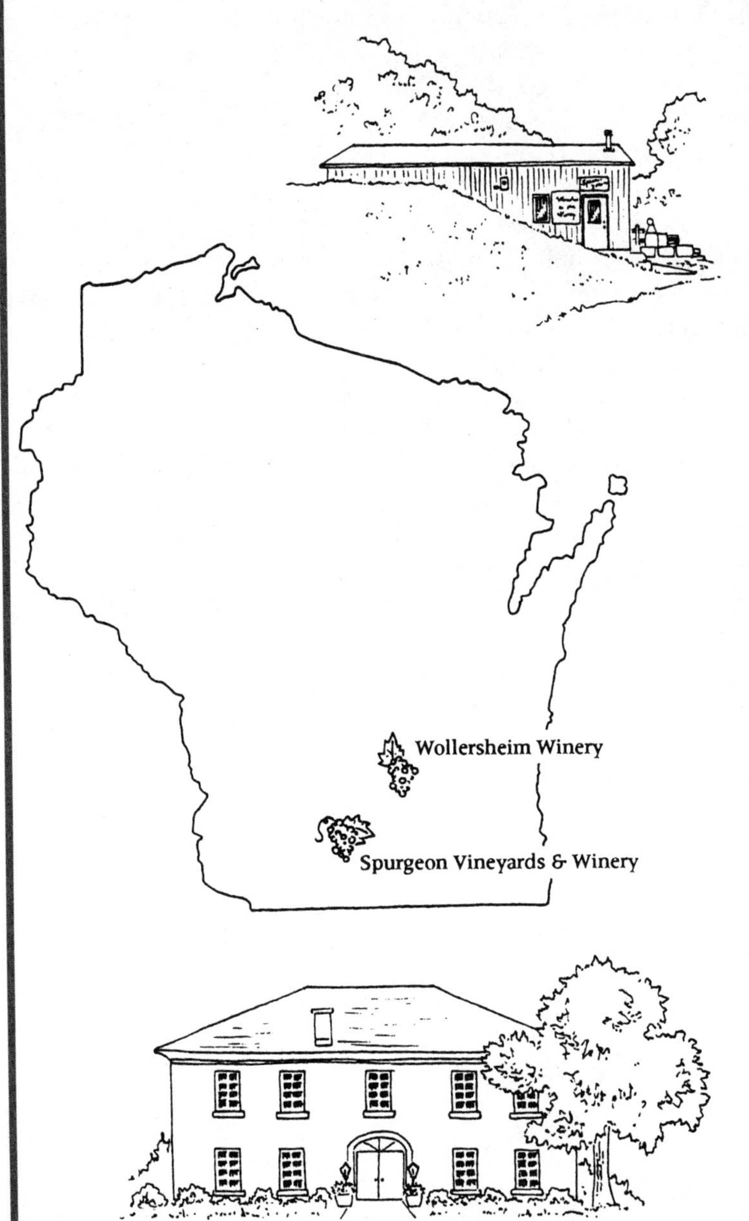

About the only way a vineyard will survive in Wisconsin is if it is located in a *microclimate*—that is, a small area where the climate differs from the larger climate around it. Microclimates are usually formed because of some type of protection or moderating force, such as a large lake, a river, or hills. It's this protection that allows the grapevines to survive.

Wisconsin horticulturist Elmer Swenson is not content with planting grapes only in the state's microclimates. He is working at breeding new grape varieties that will survive the cold weather and will ripen in the short growing season. Some of his new grapes have been planted in Wisconsin and Minnesota. Vineyard owners hope Swenson's new varieties will provide them with grapes that will survive even the coldest winters.

It was the winters that ended the early efforts of growing wine grapes in Wisconsin. Two separate attempts at planting vineyards in the 1800s failed due to the weather. Both vineyards were located on the land that is now the Wollersheim Winery in Sauk City. But Wollersheim Winery has found ways to protect its vineyards, and its attempts at bringing wine grapes back to Wisconsin look promising.

Spurgeon Vineyards & Winery

R.R. 1, Box 201
Highland, WI 53543; (608) 929–7692

"We had eighty acres of nothing," says Mary Spurgeon of the farm she and her husband, Glen, bought in 1977. "We planted our first big planting in 1978. We had to wait until 1983 until they produced enough."

They opened the winery in 1982, buying fruit to make wine. Since 1983 the Spurgeons have used only their own grapes. Mary boasts that Spurgeon is the only winery in the state making wine exclusively from its own grapes.

The Spurgeons' vineyards are perched on two hills that surround the winery. "You don't see the grapes because they're up on the two hills for temperature. It can be a ten-degrees temperature difference. And when you're talking marginal grapes, that's important," says Mary. "We have to be very careful of what varieties we're growing."

They grow more than ten grape varieties of Native American and French-American hybrids, with several small plantings of varieties in the experimental stage. The Spurgeons make a Harvest Red, Harvest White, and a Harvest Blush for $5.95. Of the Blush, Mary says, "That's our most popular one." They also make a White Wisconsin Champagne from Delaware grapes for $8.95. The Spurgeons focus on making wine that's ready to drink immediately versus wine you have to age for years.

On the thirty-minute tour you'll be able to see the winery the Spurgeons built themselves—"Except for the concrete, we contracted that out." They built the winery almost entirely underground to keep it cool during the summer. On the tour that Mary calls "one of the best in the nation," she begins with a discussion of vines. She's tied a grapevine to a fermentation tank for a prop. "You'll notice the shaggy bark? That's two years old." And she'll show you how they prune the vines each year. Mary also discusses why they don't use a mechanical harvester. "Your machine takes everything. The quality is there when it's handpicked." So the Spurgeons, even though it's labor-intensive, pick their grapes by hand each fall.

As you walk through the winery, Mary explains fermentation and filtering. She goes out of her way to put things into terms everyone can understand. Take the wine filter, for instance. Instead of talking in microns, the size of particles the machine filters out, Mary says, "It is the equivalent in size to splitting a hair fifty-two times. In other words, the very fine stuff is filtered out." She'll explain the bottling equipment, and then let each child on the tour cork a bottle. "They love it," Mary says.

It's all just Mary's way of saying that when you visit this winery, children are welcome to come and learn about winemaking too.

DIRECTIONS: From Highway 80 in Highland, turn left at the Post Office onto Highway Q. Drive for 4 miles. Turn right at Pine Tree and drive for 1.5 miles. At the intersection of Big Springs, bear left; winery is on the right.

HOURS: Tours and tastings offered daily April through October from 10 A.M. to 5 P.M.; Saturday and Sunday November through March from 10 A.M. until 5 P.M. You should call ahead during December, January, and February. Tours cost $2 for adults and 50 cents for children. Winery closed New Year's Day, Easter, Thanksgiving, and Christmas.

EXTRAS: Winery sells wine and wine accessories.

Wollersheim Winery

7876 Highway 188
Prairie du Sac, WI 53578; (608) 643–6515

One hundred and twenty-five years after vines were first planted on the site, JoAnn and Bob Wollersheim opened a winery. "We opened in 1972," says JoAnn. The man who first started planting vines on the land was the Hungarian Count Agoston Haraszthy. Arriving in the area in 1847, he planted European vines and watched his vineyard fail for two consecutive winters. He moved on to California where his work in vineyards there earned him the title of "the Father of California Winemaking."

Peter Kehl next planted vineyards on the site in 1856. His family had made wine in Germany since the 1500s. Peter planted winter-hardy American varieties and Riesling, which he buried each winter to protect the vines. When Peter died in 1871, his son Jacob began taking care of the vineyard and making wine. His wines were sold in hotels in Milwaukee and Chicago and to churches for altarwine.

The brutal winter of 1898–1899 saw the death of both the vineyards and Jacob. With their deaths, the Kehl family quit making wine. "That was the last time grapes were grown here," says JoAnn. "The next generation just didn't really care about making wine."

The Wollersheims bought the farm from the Kehl family in 1972. And what started as just a hobby, became the family business. They are joined at the winery by their son-in-law and winemaker, Philippe Coquard. The winery makes wine with grapes from the French-American hybrids in its own vineyard and from grapes from other states' vineyards.

Wollersheim makes dry reds such as Pinot Noir, Ruby Nouveau, and Domaine du Sac. Some of the dry whites are Dry Riesling and Sugarloaf White. "We specialize in dry wines," says JoAnn. The semidrys are Blushing Rosé, Prairie Fumé, and Johannisberg Riesling. Wollersheim also makes sweet and nonalcoholic wines. "We make a variety so you'll find something you like," says JoAnn. The award-winning wines sell for $6 to $12.

You'll taste the wines in what was the Kehl's carriage house. Under the carriage house are the three wine cellars that you'll see on the tour. The cellars are filled with oak barrels made from French, Yugoslavian, and Wisconsin oak. At the back of one of the cellars are 500-gallon wooden casks made by Confederate prisoners during the Civil War. The barrels were brought from Ohio in 1973.

All the walls of the cellar, along with the walls of the main house, are made from limestone 2 feet thick. The tour will also take you back behind the winery to look at the original wine cellar. The wine was stored in a cave in the hillside. It was not only a wine cave, but the place the Kehl family lived until their home was built.

Also on the tour you'll be shown some of the vineyards as well as a wonderful view of the Wisconsin River Valley. You'll see a video about how the Wollersheims make wine and then return to the tasting room for some samples. Look closely at the sales counter. It once was a 3,050-gallon wine cask.

As you leave the winery, you'll see the 150-year-old oak tree. When the Wollersheims tried to determine its age they had a sample of its rings removed. They found that the tree had purple rings. After further investigation they found that the purple rings corresponded with the years that the Kehls made wine. Evidently the

Kehls threw the skins and seeds next to the tree and the tree absorbed the color and wine. The Wollersheims think it's what gives the tree its crooked angle. It only stands to reason that at a winery that's listed on the National Register of Historic Places, even the trees should have historic significance.

DIRECTIONS: From Milwaukee or Chicago, drive north on Interstate 90/94 to the Highway 19 exit. Drive west on Highway 19 past Waunakee and then turn right on U.S. Highway 12. Drive west and just before entering Sauk City, turn right on Highway 188. Drive 2.2 miles; winery is on the right.

HOURS: Tours and tastings offered daily from 10 A.M. until 5 P.M. Winery closed New Year's Day, Easter, Thanksgiving, and Christmas.

EXTRAS: Winery sells wine and wine accessories.

Wine Vocabulary

Knowing or not knowing wine vocabulary is a lot like the difference between driving a Volkswagen or a Mercedes. Both cars will get you where you want to go, but you might enjoy the ride better in a Mercedes. You can enjoy wine without the ability to label things. If you want to be able to talk like a connoisseur, however, read on.

Acidic (tart, sour): You won't find a wine that doesn't contain some acid. The right combination of acid and sugar gives the wine balance. Acidic wines may be imbalanced, or young and sharp. In general young wines taste more acidic than older ones.

Aftertaste: Take a swallow of wine, and the aftertaste is the flavor that stays in your mouth. Also known as the "finish."

Age: Most people believe the more age a wine has the better it is. This is not necessarily true. Most wines made today are made for drinking when they're "young." A white wine made with a fruity taste is better in the first several years because after that it loses some of its fresh fruity taste. Each wine is different, so in general the rule would be to ask if the wine was made to be aged.

Appellation: A viticultural area designated by the federal government. The government delineates certain grape-growing regions because they have specific geographic features that set them apart from other areas.

Aroma: Many factors play into the "smell" of a wine. Aroma comes from the grape variety used. The aroma may have the smell of fruits, flowers, or spices. Some wines will and should have a stronger aroma, while others are more subtle. Aroma differs from bouquet. See *Bouquet.*

Astringent: An astringent wine, because of the high amount of tannin it contains, will pucker your mouth. With a well-made wine, the tannin will lessen as the wine ages. See *Tannin.*

Balance: The winemaker must find harmony among the ingredients in the wine—tannin content, sugars, acids, and alcohols. When that harmony is found, the wine has balance.

Bitter: One of the basic components of taste, but it should not be an over-powering one. A wine with a high tannin content will be bitter.

Blend: Almost all wine has some blending. This can include the blending of vintages, vats, or grape varieties. Blending is done to create a better wine.

Body: The combination of the wine's characteristics that gives the wine its weight. The actual heaviness or lightness that the wine has in your mouth.

Bottle Age: The time a wine spends in a bottle vs. the time spent in a cask or barrel.

Bottled By: Indicates the wine was produced elsewhere, and the winery named on the label was involved only in the bottling.

Bottle Sickness: Basically, the trauma the wine suffers after being bottled. It can cause the wine's flavor to suffer for weeks or months until it has a chance to recover. May also occur from jostling the bottle during shipping.

Bouquet: The wine receives its bouquet from the fermentation process, the cellaring, the aging, and anything else other than the type of grape used. With some wines you will be amazed at what different smells your nose can distinguish. Not the same as *Aroma*.

Breathing: Many people will tell you that one of the joys of wine is that it is a living thing—it changes, grows, ages, and also breathes. When wine "breathes" air, the oxygen helps release the bouquet and aroma. When wine drinkers swirl the wine around in their glasses, they are helping the wine breathe.

Brut: Term used for very dry champagnes.

Capsule: Cover or hood over the cork and bottle opening. Traditionally made of lead, with its malleability, it is being replaced by plastic.

Character: A favorable term for wine with personality and individuality. A wine that lets you know what it is without being too pushy about it.

Chewy: Much as the word implies, a chewy wine has a full body and a weight to it. Usually found in wines with high tannin content such as Cabernets.

Clearing the Palate (Cleansing the Palate): This helps you get your mouth in a condition to receive the full effect of the wine. Chewing on a small piece of bread or cracker will clear your mouth.

Cloudy: A visible sign that the wine has a problem, possibly from the sediment sometimes found at the bottom of the wine bottle.

Corky: A musty odor or taste due to the flavor of the cork being passed to the wine.

Crush: Generally speaking, crushing the grapes is part of making wine. Specifically, people refer to harvest time as the crush.

Decant: Decanting a wine involves pouring it from the bottle into a decanter. As you finish pouring out the wine, you should be careful not to pour the sediment that rests at the bottom of the bottle into the decanter. Decanting wine achieves two goals. It removes the sediment from the wine, and it gives the wine a chance to breathe.

Disgorge: The process used to eliminate the sediment in champagne. The solids are gathered in the neck of the bottle. Then the neck is frozen, and the solid plug of sediment is removed.

Dry: A wine without a strong sweet flavor. A dry wine will contain little, if any, sugar.

Enology: The study of wine and winemaking.

Estate-bottled: Traditionally estate-bottled wine included only wine made from grapes grown, harvested, fermented, and bottled by the same person or winery. Recently it has been used for wine produced from different vineyards but under the control of one owner or within the same viticultural area.

Fermentation: All great wines, and bad wines, begin and end with a simple chemical reaction: yeasts consume the sugar found in grapes, and the byproducts are alcohol and carbon dioxide.

Filtering: A process to remove yeast and other solids from wine after it is fermented.

Fining: A process that removes yeasts, solids, and sediments from wine after it has been fermented. Traditionally egg whites were added to the top of the container, and as they settled, the egg whites carried down particles that had been suspended in the wine. The process clarifies the wine.

Finish: The taste that settles and lingers in your mouth after you have swallowed the wine. The sensation can linger or leave immediately, and it can be strong or weak.

Flowery: Simply enough, a bouquet that reminds you of flowers. Some wines have it, and they should. Others have it, and they shouldn't. It depends on the variety.

Fortified: Wines, such as sherries and ports, that have extra alcohol added during production.

Foxy: The term originates from the American Fox grape, which grows wild in the Northeast. Now it includes the odor and taste of wine made from native American grapes such as Concords or Muscadines.

Fruity: A bouquet that reminds you not necessarily of one particular fruit, but of fruits in general.

Full Bodied: A wine with a high level of alcohol usually feels full bodied. The wine will feel dense and weighty in the mouth.

Generic: A wine with an alias—that is, a wine with a name that formerly represented a specific area and which gained international recognition and is now used by others, such as a Chablis or a Burgundy.

Green: A young acidic wine that can usually improve with more aging.

Hybrid: A grape variety developed by crossing two or more grape varieties. French-American hybrids, a cross of vinifera and native American vines, are used in the Eastern and Central United States because of their hardiness and resiliency. Examples are Seyval and Vidal.

Labrusca: Native American grapes such as Delaware, Concord, and Catawba. Said to have a distinct foxy taste.

Late Harvest: A wine made from ripe grapes and gathered after the usual harvest to increase the sweetness or the amount of alcohol.

Lees: After wine has fermented and aged in a cask and then been drawn off, lees are the sediments remaining.

Legs: When you swirl your wine in the glass, rivulets that run down the *inside* of your glass are called legs. Legs can indicate a rich, full-bodied wine.

Made and Bottled By . . . : At least 10 percent of the wine in the bottle was produced by the winery listed on the label. That leaves up to 90 percent to have been bought in bulk elsewhere.

Méthode Champenoise: The traditional method of making sparkling wine, with the wine fermented in the bottle, thus forcing bubbles of carbon dioxide into the wine.

Muscadine: The *rotundifolia* species of grape. These native grapes, such as the Scuppernong, grow in clusters instead of bunches. Native to the southern United States.

Must: The grape juice before fermentation. Not to be confused with *Musty.*

Musty: An unpleasant odor or flavor that could have been caused by moldy casks or moldy corks.

Nose: A nose includes all aspects of the smell of a wine, such as aroma and bouquet.

Oaky: A flavor or aroma lent to the wine from aging in oak barrels or from oak chips. Some winemakers go to great lengths to find just the right oak to flavor their wines.

Oxidized: Wine reacting with oxygen to such an extent that it is detrimental to the wine, causing it to turn a brownish color and taste flat.

Produced and Bottled By . . . : The winery listed on the label made at least 75 percent of the wine.

Racking: Transferring wine from one container to another one. Racking clarifies a wine because the lees are left at the bottom of the first container.

Residual Sugar: The amount of sugar remaining in the wine after the fermentation process has been halted. Residual sugar will determine the degree of sweetness that the wine has.

Sec: This describes a semidry still or sparkling wine.

Sediment: The solid matter that precipitates out of the wine.

Sheets: Sheets are wide rivulets of wine clinging to the sides of your glass after you have swirled your wineglass. Sheets are larger than "legs." Sheets can indicate a rich full-bodied wine. See *Legs.*

Sparkling: A sparkling wine contains bubbles of carbon dioxide gas. The gas may be pumped into the wine, or it may occur naturally in the sparkling wine if it was made in the *méthode champenoise.* Champagne is a sparkling wine. The French believe that only sparkling wine from the Champagne region of France should be called champagne.

Stemmy: A wine with the flavor of stems vs. a fruity flavor. It can be caused by fermenting the wine in contact with the stems or by crushing the grapes so violently that the tannin flavor of the stems and seeds is released.

Sweet: A taste dependent on, but not necessarily limited to, the amount of residual sugar in the wine. Other factors such as alcohol and glycerin content can give the wine a "sweet" taste.

Tannin: Chew on a grape seed, skin, or stem, and you'll taste tannin. Tannin can also be found in oak casks. It is a natural preservative and plays an important part in aging wine. Tannin has a bitter taste and is most noted in young wines, but it mellows after a few years. It is a necessary ingredient in most reds and a few whites.

Texture: Texture is the feel of the wine as opposed to its flavor. It is the impression of tiny particles in your mouth.

Varietal: A wine named for the single variety or the primary grape variety used. The wine should have the pronounced aroma and flavor for which that grape variety is known.

Vat: A large container or tank used for fermenting or blending wine. Vats can be made from wood, concrete, or stainless steel.

Viniculture: The science and study of grape production.

Vinifera: The name for the species of European grapes. Examples are Chardonnay, Pinot Noir, and Cabernet Savignon.

Viticulture: The science and art of growing grapes.

Woody: The smell and taste of the wood in which the wine was aged. The term can also be applied to a wine in a derogatory sense when the wood aging has overpowered the wine or when the wood flavor does not complement the grape variety used.

Yeast: Microorganisms that chemically react with sugar in a process called fermentation. During fermentation the yeast eats the sugar and produces alcohol and carbon dioxide. Wild yeasts are found naturally on grape skins, though most winemakers today use special yeasts for fermentation.

Yeasty: A flavor yeast cells can give wine. Also, the odor of yeast.

Index to Wineries

A

B

C

D

F

G

H

I

K

L

M

N

O

P

R

S

ABOUT THE AUTHOR

Destined to write a book about wine, Pamela Stovall grew up in a house located between Grape Street and Vineyard Avenue in her hometown of Grand Rapids, Michigan. She has spent a considerable amount of time studying wine while living in Texas, Alaska, Florida, and Michigan and while serving as a Peace Corps volunteer in Ecuador, where she became familiar with South American wines.

A freelance writer, Stovall's stories have been published in numerous magazines and newspapers such as *The Dallas Times Herald, Accent Magazine, The Houston Post, Woman's World, Association Trends, The Professional Communicator, The Atlanta Publicity Handbook,* the *Austin American-Statesman, The Chattanooga News-Free Press, The Grand Rapids Press, Careers in Communications,* and the *Kodiak Daily Mirror.*

For years she has found her weekend visits to wineries entertaining and informative, and she enjoys sharing her adventures with like-minded readers.